T0280405

AIRCRAFT INTERIOR COMFORT AND DESIGN

Ergonomics Design and Management: Theory and Applications

Series Editor
Waldemar Karwowski
Industrial Engineering and Management Systems
University of Central Florida (UCF) – Orlando, Florida

Published Titles

Aircraft Interior Comfort and Design
P. Vink, and K. Brauer

Ergonomics in Developing Regions: Needs and Applications
P. A. Scott

Ergonomics and Psychology: Developments in Theory and Practice
O. Y. Chebykin, G. Z. Bedny, and W. Karwowski

Human–Computer Interaction and Operators' Performance: Optimizing Work
Design with Activity Theory
G. Z. Bedny and W. Karwowski

Trust Management in Virtual Organizations: A Human Factors Perspective
W. M. Grudzewski, I. K. Hejduk, A. Sankowska, and M. Wańtuchowicz

Forthcoming Titles

Ergonomics: Foundational Principles, Applications and Technologies
P. McCauley-Bush

Knowledge Service Engineering Handbook
J. Kantola and W. Karwowski

Handbook of Human Factors in Consumer Product Design, 2 vol. set
W. Karwowski, M. M. Soares, and N. A. Stanton

> Human Factors Interaction: Theories in Consumer Product Design
> Human Factors Design: Case Studies in Consumer Product Design

Organizational Resource Management: Theories, Methodologies, and Applications
J. Kantola

Neuroadaptive Systems: Theory and Applications
M. Fafrowicz, T. Marek, W. Karwowski, D. Schmorrow

Manual Lifting: Guide to Study of Simple and Complex Lifting
D. Colombini, E. Ochipinti, E. Alvarez-Casado, and T. Waters

AIRCRAFT INTERIOR COMFORT AND DESIGN

PETER VINK AND KLAUS BRAUER

CRC Press
Taylor & Francis Group
Boca Raton   London   New York

CRC Press is an imprint of the
Taylor & Francis Group, an **informa** business

CRC Press
Taylor & Francis Group
6000 Broken Sound Parkway NW, Suite 300
Boca Raton, FL 33487-2742

© 2011 by Taylor and Francis Group, LLC
CRC Press is an imprint of Taylor & Francis Group, an Informa business

International Standard Book Number: 978-1-4398-6305-3 (Paperback)

Library of Congress Cataloging-in-Publication Data

Aircraft Interior Comfort and Design / P. Vink.
    pages cm. -- (Ergonomics Design and Management: Theory and Applications ; 5)
    Includes bibliographical references and index.
    ISBN 978-1-4398-6305-3 (pbk.)
    1. Aircraft cabins--Human factors. 2. Interior decoration--Human factors. 3. Air travel--Public opinion. I. Vink, P. (Peter), editor.

TL681.C3A38 2011
629.134'42--dc22   2010048436

Visit the Taylor & Francis Web site at
http://www.taylorandfrancis.com

and the CRC Press Web site at
http://www.crcpress.com

Contents

Foreword

There is considerable research and development going on in the industry that deals with aircraft interiors, which is offering many opportunities for improvement in this field. Much of the work in this area is not published for public consumption, even when sharing knowledge on passengers' demands could further help the industry.

This book consists of public information on the latest understanding of comfort, on what 10,032 customers like and dislike, and what other studies report about customer opinions. All of this translates into aspects of aircraft interiors, which can be an inspirational source of information for airlines, aircraft manufacturers, their suppliers, researchers, and designers in this field.

It is interesting to discover that the airline industry performed very well with newer planes, receiving a higher average comfort rating than older aircraft. This shows that attention to the interior pays off and *Aircraft Interior Comfort and Design* could be an inspirational source to continue the improvement throughout the industry.

Antje Terno
Dipl.-Ing., Manager Cabin Seat Development/Change Leader
Airbus Operations GmbH

About the Authors

Peter Vink, PhD, is a specialist in the field of comfort and design. Dr. Vink has written more than 250 papers and 8 books on comfort and design, and assists many companies with his knowledge and expertise on the subject. Since 1998, he has been the head of the Ergonomics and Innovation Department at TNO (Netherlands Organization for Applied Scientific Research) and, since 2001, he also has been a professor at the Delft University of Technology on the faculty of Industrial Design Engineering where he guides MSc students in designing comfortable products. Dr. Vink also guides 10 PhD students who are active in the field of interior design.

Klaus Brauer has been active in the fields of aircraft interiors and passenger comfort for two decades. He was involved in the design of the interior of several Boeing aircraft, including a leading role in the design of the 787 Dreamliner interior. He retired from his role as director of Passenger Satisfaction and Revenue at Boeing Commercial Airplanes in 2009. He is currently a consultant to B/E Aerospace.

1 What Every Manufacturer and Airline Should Know about Comfort

Overview: The suppliers of aircraft and interior furnishings to the airline industry have made big steps forward that have resulted in significantly higher passenger ratings for newer aircraft compared with older types still in service. Airline sales are driven to some degree by comfort and, as a result, improved comfort provides an opportunity to attract more passengers. However, having reached a high level of comfort, it is now more difficult to improve comfort further still. This book is intended to help identify new opportunities for comfort improvement in the different phases of the passenger experience:

- Setting expectations
- Comfort at first sight
- Short-term comfort
- Long-term comfort

In each phase, the input to the passengers' senses can be optimized, which leads to new opportunities for comfort improvement. It is important to recognize that not only hardware can improve comfort, but recent research has shown that passenger expectations, crew professionalism, and final design play major roles in the comfort experience.

TWO COMFORT STORIES

Being a comfort nerd, I asked the airline passenger seated next to me on a recent flight how comfortable he was after sitting for four hours in the economy seat. He answered, "After desperate attempts to check in online, I had to line up 30 minutes for check-in and 15 minutes for security. Four hours in this airplane is terrible. There is no information on where we are in space, no free meal, no free water, no blanket, no pillow, no movies, no games, and no radio. Did you speak to one of the four air hostesses who all seem to have forgotten their course on 'being friendly'? The four hours in the narrow space seemed like an eternity, and my body (especially knees and shoulders) are protesting."

In the survey described in Chapter 3 with 10,032 passengers, we did discover many similar stories, but we also received neutral stories and many stories that are quite positive. One of the stories associated with a very high comfort score read: "Wow, I could check in online by only four clicks for a seat at the emergency exit row and I got a bonus route description to the gate. A warm welcome by the crew, good food, a few drinks, and space to work with a laptop. It was more than I expected. Having your elbows in front of your body is not the most comfortable position, of course, but it works. Before landing, we got a nice movie on the town where we were landing. It makes the three hours fly by. The walking distance to the railway station was really short, giving me extra time."

These stories were not selected at random, but they do clearly speak to the importance of several aspects of the passenger experience, which a study of 10,032 passengers proves to be vitally important to the perception of comfort. The first impression of the environment is very important, as are expectations and emotional considerations. Also, short-term and long-term physical comfort does, of course, play a major role in travellers' overall sense of comfort.

ROLE OF COMFORT IN SALES

One airline strategy that is aimed at selling more tickets is to provide a superior comfort experience. This is not easy because it requires the coordinated effort of many individuals in the airline business:

- Management that understands comfort from the perspective of the passenger and that manages airline resources to deliver comfort in a profit-maximizing way.
- Purchasers who calculate the cost-benefit trade for equipment investments required to deliver superior comfort.
- Staff, especially the staff that is in close contact with the client: designers of the Web site, individuals who answer complaints, ground staff, pilots, and flight attendants.
- Aircraft manufacturers that have and execute the vision that the airplane should be designed around the passenger experience.

- Aircraft interior component manufacturers, designing in-flight entertainment, lighting, seats, carpets, and other interior parts.
- Cleaning and maintenance companies that make the look and function of the plane optimal.

In principle, airlines can increase their profit margin by reducing maintenance costs. However, according to Brauer (2004), at a typical airline, a 14 percent reduction in maintenance costs will result in only a one percentage point improvement in the airline's profit margin, while a passenger revenue increase of only 1 percent has the same result.

To increase passenger revenue, we need to understand the flight selection behaviour of passengers. According to Brauer, most passengers first select the most convenient route and departure time at the best price. In those cases in which the passenger is indifferent between equally convenient flights at a similar price, other aspects break the tie. These other aspects include comfort, service, the airline's reputation for on-time performance, and marketing programs, such as frequent flyer programmes. For short distances, on-time performance is more important and, for long-haul flights, the comfort and service aspects play the most important roles. Under the foregoing flight selection paradigm, individual passengers never make a choice to pay more for more comfort; however, the revenue management system of the preferred airline does react to the resulting increase in demand by reducing the number of seats sold at discount fares and, as a result, capture somewhat higher loads and yields.

While it is less common than the schedule and price-driven paradigm described above, it is true that some passengers, in fact, do choose a slightly less convenient flight or a slightly higher fare to fly aboard their favourite airline. Reasons for having a favourite airline are many; where the reason is comfort, the value of comfort, in terms of tickets sold, is clear.

THE DIFFICULTY OF MAKING PEOPLE FEEL COMFORTABLE

There is another element complicating the effort to make airplanes more comfortable: The fact that each passenger decides whether or not he or she is comfortable. There is at least one thing on which passengers have more knowledge than do airplane manufacturers, flight attendants, and airline managers, which is the passenger's own sense of comfort. We cannot answer the question of whether a particular passenger feels comfortable in an airplane. The passenger is the only one who can do that. That is one of the reasons why it is so difficult to design for comfort or to run a company where comfort is an important determinate of success. Every passenger has his or her own subjective sense of comfort.

THE GOOD NEWS: IT IS POSSIBLE TO MAKE PEOPLE FEEL MORE COMFORTABLE

The study of 10,032 passengers reported in Chapter 3 shows that newer aircraft are regarded significantly better with respect to comfort than are older ones. These

TABLE 1.1
Opportunities to Influence Comfort

Comfort Process Phase	Opportunities
Expectations	Optimize brochures, Web sites, check-in system, seat choice
First sight	Nice entry, good looking interior, spacious seating place
Short-term comfort	Positive attention of crew, a personal gift
Short-term discomfort	Seat feels good, no obstacles, no pressure, no stress on the body
Long-term comfort	Unexpected positive attention, popular movies, good view, opportunities for the passengers to do their activities
Long-term discomfort	Variation in posture possible, good form, and cushioning of the seat
Restore or affirm	Tell that the bad experience was an anomaly, offer a possibility to complain, or affirm the good experience

findings should give those involved in the industry a sense of satisfaction with regard to the job they have done in improving comfort. They have shown that it is possible to improve the comfort experience. Bringing comfort to an even higher level will be more difficult when starting from this new higher standard. However, increasing the comfort level further is still possible, as there is still more knowledge to bring to bear in making passengers feels comfortable.

The insights presented in this book may be new to the reader or they may only serve to reaffirm the reader's own perceptions and intuition. In either event, when applied in the daily work of managing the passenger experience, they can lead to further improvements. This will not be easy, but it is possible. To make that work easier, it may be helpful to distinguish different moments influencing the comfort experience. Each of these moments or phases is addressed by unique service and design requirements.

Based on years of comfort research (Vink, 2005), we made a distinction in different comfort experiences (Table 1.1):

- Building up the expectations
- The first sight comfort
- Short-term discomfort
- Short-term comfort
- Long-term discomfort
- Long-term comfort

It is important to distinguish these phases in the comfort experience from one another because improving each phase requires a different approach and focusing on only one phase is not sufficient. If you increase long-term comfort and discomfort, but it is not to be seen in brochures, Web sites, or the moment you enter the plane, the improvements will not influence sales as effectively as when the improvements are reflected in all relevant phases of the comfort process. On the other hand, if there is a very good brochure and Web site, the expectations are high and the flight may prove to be a disappointment. In a study by Blok, Vink, and Kamp (2007), it appeared that the comfort scores of business class passengers were not significantly different from the scores of economy class passengers. On a scale from 0 to 10, 10 being maximum comfort, both scored around 7. Therefore, a reasonable explanation is that business class passengers expect more and, therefore, are more critical, resulting in approximately the same comfort scores for economy and business class. These expectations are very important. If we look to passengers who had a terrible experience in their previous flight, their comfort score was significantly higher on the subsequent flight than that reported by passengers who did not mention their previous flight. It is not surprising that, when a passenger is upgraded because of the lack of seats in the economy class, their scores are significantly higher. The importance of expectations means that the manufacturers, airlines, and crew must be aware of the experience provided in other airplanes and by other airlines when they deal with frequent fliers because passengers expect the same as in their previous flight if they pay the same price.

COMFORT THEORY

You might be bored by seeing the title of this section. This means that your *expectation* will be low regarding the fun element in this section. You expect that the text will be difficult and far from exciting. This influences the way you read this. The same principle is true for comfort. In Table 1.1, part one of the comfort theory is shown, the phases that have an important influence in the comfort experience. It starts with the expectations. The brochures, Web site, previous experiences, stories of friends, they all set the expectation and thereby influence the comfort experience.

The moment you see the interior of the airplane you are having your first impression experience. This first sight does influence the comfort experience during the flight as well. This impression could be different from the one while seated, but it does have an influence. In an experiment where new BMW automobile seats were tested, one of the seats had blocks with several colours (Bubb, 2008). It simply did not look nice. The comfort was also rated lower during a first sight, while the form was exactly the same as a seat that was black. Even during sitting, the comfort was rated lower in the beginning. This may sound strange because the moment you sit on the seat you do not see the blocks of colours anymore. The first sight seems to be of importance throughout the total comfort experience.

When considering your comfort as you read this, several thoughts may occur to you. You may experience *discomfort* in your room due to a draught or because it is too hot. You may feel pressure points on your bottom. Perhaps your bottom does not fit in the seat or maybe you do not have an appropriate lumbar support.

You may feel *comfortable* because you have very good company or are in a luxurious hotel room. You may feel comfortable due to this engaging text and my pleasant words and attitude trying to comfort you. Or you may feel thrilled by my enthusiasm on the topic because it is really a possibility to attract new passengers, sell products, or a way to make people feel at home in the airplane.

Usually, you do not think about comfort at all. You experience *no discomfort*. The chance that this will happen reading this book is small because I am making you aware of the concept of comfort.

Perhaps you notice that a distinction has been made between discomfort and comfort. There are studies in the literature that have shown that while sitting discomfort is more related to pressure points and stiffness. Comfort, on the other hand, is more connected to luxury and refreshment (see, e.g., Helander and Zhang, 1997). This distinction is made in this book as well. Both experiences can be found after just being installed in your seat (*short term*) and after a few hours (*long term*). It is of importance to make this distinction as it requires other activities of designers, management, and crew.

COMFORT MANIFESTATIONS

Thus, comfort could have three manifestations: (1) discomfort, (2) comfort or comfortable, and (3) nothing is experienced or no discomfort. I will put these three appearances into a theoretical framework. The motivation theory of Herzberg, Mausner, and Snyderman (1959) is described first because of the similarity of this framework with this motivation theory. In the late 1950s, Herzberg was considered by many to be a pioneer in motivation theory. He interviewed employees to find out what made them satisfied and dissatisfied on the job. Physical factors, according to Herzberg, cannot motivate employees, but can minimize dissatisfaction if handled properly. In other words, they can only dissatisfy if they are not all right. Dissatisfaction is related to company policies and salary (Table 1.2). Motivators, on the other hand, create satisfaction by fulfilling individuals' needs for meaning and personal growth. These are issues such as the work itself and advancement and are related to satisfaction.

In comfort, a similar division can be made. Absence of discomfort does not automatically result in comfort. Comfort will be felt when more is experienced than expected. This is supported by research of Zhang, Helander, and Drury (1996) and Helander and Zhang (1997). Based on questionnaires, they found that discomfort is more related to physical characteristics of the environment, like posture, stiffness, and fatigue (Table 1.3). In the case of absence of discomfort, nothing is experienced. To notice comfort more should be experienced. Comfort is related to luxury, relaxation, or refreshment.

TABLE 1.2

Factors Influencing Satisfaction or Dissatisfaction

Dissatisfaction	Satisfaction
Company policies	The work itself
Administrative procedures	Achievement
Salary	Recognition
Working conditions	Advancement

Source: Adapted from Herzberg, F., B. Mausner. and B. B. Snyderman. 1959. *The motivation to work.* Somerset, NJ: Transaction Publishers.

TABLE 1.3
Factors Influencing Comfort or
Discomfort during Sitting

Discomfort	Comfort
Fatigue	Luxury
Pain	Safe
Posture	Refreshment
Stiffness	Well-being
Heavy legs	Relaxation

Source: Adapted from Zhang, L., M. G. Helander, and C. G. Drury. 1996. *Human Factors* 38 (3): 377–389.

Therefore, three conditions of comfort, in fact, can be distinguished:

- Discomfort: The participant experiences discomfort because of physical disturbances in the environment.
- No discomfort: The participant is not aware of the fact that there is no discomfort.
- Comfort: The participant experiences noticeably more comfort than expected and feels comfortable.

Also, five moments in time are influential and should be taken into account when trying to optimize the comfort experience: expectations, first sight, short term, long term, and "after comfort service." In fact, opportunities to reduce discomfort or improve comfort exist in each of these five moments. Examples of influencing discomfort and comfort are provided in the next section.

INPUTS LEADING TO (DIS)COMFORT

The comfort schema describing the inputs leading to the output (dis)comfort is visualized in Figure 1.1. On the right side in the figure, we see the output: comfort, no discomfort, and discomfort. This output, the experience of discomfort or comfort, is partly due to ourselves, our history of comfort experiences, and our current state, which could be excited or relaxed. The experience of comfort and discomfort also is caused by external stimuli (input). The inputs are shown on the left in the figure. To illustrate this with an example: Our sensors receive the pressure. After this input, the selection and weighing processes begin. Our state of arousal and past experiences influence these weighing processes, and based on these processes, the product causes comfort, discomfort, or nothing.

In this section, each element of the model influencing (dis)comfort is described separately. In reality, these elements are not separate. It is not known precisely how the elements are related to each other and what the contribution of each element is to the total

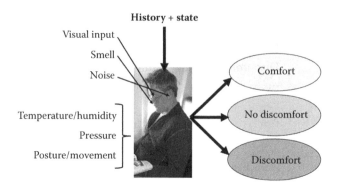

FIGURE 1.1   The comfort input/output schema. The feeling of (dis)comfort is determined by the input recorded by sensors and the information processing that is influenced by the history and state of the participant.

experience. That is why it is important to perform experiments with products in the design phase in an environment as close to the natural setting as possible and test every product or approach. In Chapter 3, an attempt has been made to define the influence of different phases in the flight experience on comfort and then to weight these influences.

HISTORY

History influences the experience. This is important for product designers and the crew. Interiors should have at least the level of comfort people are used to. A nonadjustable office seat will be experienced as not comfortable in the Netherlands because almost every Dutch office worker is used to this. A caveman will not have this problem. Of course, this is only a hypothesis because the opinion of the caveman is now difficult to verify in an experiment. We always evaluate the appearance and styling of a product with our past as a reference, and we always evaluate the service related to past experiences. That means that a product designer and crew should know the history of the target group. In business class, passengers are used to a welcome drink and a seat that is adjustable in various positions. The airlines that did not offer this service had a significant lower comfort rating in our study among 10,032 travellers compared with the other airlines. A great deal of research is now done regarding seats that adjust themselves to the most ideal comfort position. In late 2008, Dr. Zenk performed experiments with intelligent seats in a BMW automobile, and the test subjects gave them high ratings (Zenk, 2008). Another new finding is that people like to have their feet off the ground while watching a screen (Figure 1.2). In 2008, industrial designer Rosmalen et al. (2009) tested a new lounge seat based on this principle and, of course, the test subjects were very enthusiastic. If these features are available in business class in a few airlines, passengers will ask for it, and it will influence their comfort ratings.

STATE OF MIND

Our *state of mind* also influences whether we experience discomfort or comfort. After a few hours walking or running before entering the gate, your seat is probably

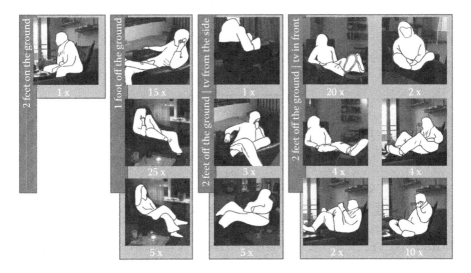

FIGURE 1.2 Rosmalen et al. (2009) found when individuals are given the freedom to choose a comfortable position while watching a screen, the feet are often off the ground. Pictured are some of the positions frequently observed.

more comfortable than after waiting for three hours in a chair at the gate. If you have an important appointment and must arrive on time, your state of mind in the airplane is also different than when you only have to go to a hotel and have more freedom regarding arrival time. This all influences your comfort experience. This also is shown in the comfort scores of the 10,032 studied trip reports. The average comfort score of all flights is 7, while the average comfort score of the passengers experiencing a delay of more than four hours is only 5.8. A rude crew influences the score even more. Passengers on flights where pilots did not give any information and when flight attendants were impolite to their passengers have a score of 2.4. Even not giving enough attention to passengers reduces the mean comfort score to 3.9. So, emotions, feelings, and mood play a role in the way someone evaluates a product.

Visual Input

The *visual input* also influences our experiences. Visual information plays a major role. It is the first impression of comfort. Humans see a shape, size, glossiness, and lightness of an object and form an impression on how comfortable it is. This visual impression is not an objective quality, but is a mental construct (Nefs, 2008). For example, objects might look flatter when they are made of a lighter material. It is important to realize that comfort is not only influenced by styling or appearance. Colour plays a role as well. Kuijt-Evers (in Bronkhorst et al., 2001) showed that 49 experienced office workers evaluated one out of four office chairs as less comfortable based on visual information. One chair was brown and the other four more freshly coloured. The brown chair was rated less comfortable, while the form and material were equal for all chairs. Contrary to what was expected, this chair was evaluated positively after using it for some time.

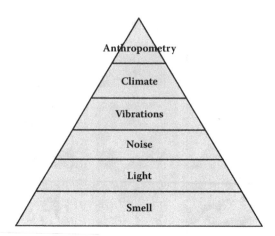

FIGURE 1.3 The discomfort pyramid based on the work of Bubb (2008). A bad smell has so much influence that it overrules all other aspects. In fact, smell, light, vibrations, noise, and climate are at a rather high standard in current airplanes. This results in the anthropometry being the focus of more attention. Attention to anthropometry is of little importance if the other aspects are not acceptable. In commercial aviation, service may merit a level above anthropometry in the Bubb discomfort pyramid.

SMELL

Smell also influences our perception of comfort. Different authors (e.g., Theimer, 1982) report that smell influences our experiences and that we are mostly not aware of this effect. It even influences our sexual activity, aggression, and territorial behaviour. We also are aware of certain smells. Odours warn us of dangers. We quickly smell spoiled food or smoke from a distant fire and become alert. According to Bubb (2008), a specialist in the field of comfort, smell is a most basic influential aspect (Figure 1.3). When your neighbour flatulates, you will involuntary move your body away to avoid the smell and feeling comfortable will be very difficult in the context of the discomfort associated with the odour. Just like other inputs, every person will react differently to the input, but a bad odour in an airplane will have a significant influence on the comfort of the majority of people. In this case, information on the odour is also very important.

Distel and Hudson (2001) showed in an experiment that odours of everyday products are experienced as more pleasant when subjects are told the name of the product causing the odour than when the source of the odour was not identified.

NOISE

Noise is a type of input that can influence comfort positively or discomfort negatively. Noise of an engine while working in an earth-moving machine can lead to discomfort (Vink, 2005), while the sound of a Harley Davidson is a kind of music to some of us. Egmond (2008) states that many people use auditory cues unconsciously. For instance, the sound of water boiling informs us about the progress of meal preparation. In an airplane, sounds can make us feel comfortable. During takeoff, it is to be expected that the engines make noise. However, lowering the power at cruising altitude in the middle of a flight could be heard by passengers, and they could feel uncomfortable if no explanation for the reduced noise level comes to mind. According to Bubb's discomfort pyramid (see Figure 1.3), it also could overrule a seat that fits well to your body (anthropometry).

Until now, inputs to the human body have been clearly located. For sight, the sensors are our eyes and, for noise, we use ears, but for some inputs the sensing organs are spread all over the body and the input is provided by a mixture of different organs. Therefore, in the model (see Figure 1.1) this input is drawn as one cluster.

TEMPERATURE AND HUMIDITY

Temperature and *humidity* also are related to (dis)comfort. Searching for comfort on the Internet leads one mostly to temperature and climate issues. Air conditioning, office temperature, draught, and humidity are often associated with comfort. A pleasant climate is often not noticed, but a high or low temperature attracts attention and discomfort is perceived. According to Carrier®, one of the largest manufacturers of indoor air quality systems, indoor air quality is the most important reason why office rental contracts are not extended (www.carrier.com). This is certainly not an objective source, but it indicates some importance. Many studies indicate that having control on your own climate influences your comfort (Lee and Brandt, 2005; Bordass and Leaman, 1997). In our survey among the 10,032 passengers, climate was not often mentioned as a problem by passengers (less than 5 percent). This suggests that current airplanes are sufficiently equipped to provide an acceptable climate. If the issue was mentioned in the trip reports, sometimes these problems were solved by the crew. Flight attendants were willing to adjust the temperature. However, in some cases, passengers had the feeling that they were not taken seriously. Their complaint was not handled or they did not get information on why the temperature was not adjusted. Dry air was also mentioned especially in long-duration flights. In those cases, passengers indicated that their eyes, nose, or mouth felt dry and made breathing less pleasant. However, climate did influence the comfort score negatively in case of a cold draft, flight attendants that were not willing to adjust the temperature (or did not mention the reason why it was not adjusted), a high temperature, dry air, or cold feet.

PRESSURE AND TOUCH

Different studies show the relationship between *pressure* and discomfort (Goossens, 1998; Goossens, Teeuw, and Snijders, 2005). To feel pressure, we have sensors located in our skin. Generally, a better distribution of pressure between the seat or handle and the human body leads to less discomfort. A literature survey (Looze, Kuijt-Evers, and Dieën, 2003) showed that, of all objective measuring methods, pressure has the most clear relationship with discomfort. In this area Goossens, Teeuw, and Snijders (2005), Mergl (2006), and Zenk (2008) have done some impressive work. Goossens and colleagues showed that participants are able to perceive small differences in pressure in their bottoms and could translate this to discomfort. Mergl made a pressure map of the human body for the ideal seat pressure distribution (see Chapter 2). This ideal pressure distribution leads to a high comfort rating. Apart from pressure, we have also *touch*. Textures of handles have an influence on the feeling of comfort. Sonneveld (2007), in her PhD thesis, describes how we can take these feelings into account during the design process.

POSTURE AND MOVEMENTS

The *posture* and *movements* determined by the product also can lead to discomfort. In the long run, discomfort could even result in musculoskeletal disorders (Hamberg, 2008). In the fourth European working conditions survey (Parent-Thirion et al., 2007), the most often reported health problems are musculoskeletal disorders (backache and muscular pains). Back pain is found in one-third of all European workers and neck/shoulder pain is found in almost one-quarter of European workers. Thus, the problem is significant enough to require attention. This presents a compelling opportunity for designers—design products that reduce musculoskeletal injuries. Establishing a reduction in discomfort in experiments during the design process is needed to prevent musculoskeletal injuries as well. If you employ a proven method of measuring local postural discomfort, you can even predict complaints (Hamberg, 2008). An opinion that is gaining support among scientists is that sitting in itself is no risk factor for back complaints. Nordin (2005) made an overview of all high quality epidemiological research studying the relationship between sitting and back pain, and she came to the conclusion that there is no evidence of a connection. There is some evidence for a relationship between back complaints and sitting in restricted postures or between back complaints and sitting in combination with vibration, but for sitting alone there is no evidence. To avoid imposing a restricted posture, it is important for airplanes to make variation in posture possible.

PERSONS INFLUENCING THE INPUT

The comfort model can help in determining where the greatest attention is needed. Aesthetics deserve attention to influence the visual input, and attention needs to be paid to odours and noise to influence perceived smell and sounds. Draft, humidity, and temperature can be optimized and the seat form can be changed to influence pressure distribution and posture. These are all fields largely manageable by designers and engineers. This, of course, is important, but the management of maintenance and cleaning also have a tremendous influence on comfort because inoperative in-flight entertainment (IFE), broken seat parts, and dirty interior components increase discomfort. A field that gets a good deal of attention is the "emotional" aspect. This is partly determined by the product. An interaction with a product in itself can present an emotional experience. There have been studies in which an awkward lamp influenced the emotions of a user (Ross, 2008), and there is a great deal of information on how emotionally appealing products can be designed (Schifferstein and Hekkert, 2007). Emotions are influenced by service as well. As described above in the study of Blok, Vink, and Kamp (2007), the influence of the crew is immense. To illustrate this, two examples of reports leading to a comfort score of 0 and 1, respectively, are given:

The ground staff at the check-in desk was rude and was talking with friends on the phone while I was trying to get on the plane in time. The in-flight service was similar. The passenger next to me became ill and I pushed the call button. No flight attendant came. I went to the front of the plane and explained that I had pushed the call button. The ladies kept on talking to each other and said without real interest: "The ringing call button can really be an annoyance to us." After convincing them there was an ill passenger, they were willing to bring some water. However, I think water was not the only thing needed in this case. Attention and asking what is going on would have been more appropriate. The pilot did give messages that I couldn't understand because he was talking too loudly into the microphone. Another passenger asked a crew member to help her complete her immigration form and she was told that the crew was "too busy." I didn't have that impression as they seemed to be chatting and laughing as a group in the front of the plane.

I attempted to use online check-in, but the system kept giving me an error message. At the airport, none of the machines would issue a boarding pass either. I had to stand in line and, after waiting 20 minutes, I explained the problem. No reaction, the person at the desk just printed the boarding pass. At the business lounge, the staff at the desk was reading the newspaper. After waiting a while, I asked for service and I was told that someone would come. After waiting longer, I was allowed to go inside. It was a mess, most of the food was gone and newspapers and dishes were everywhere. An announcement was made to go to the plane and after arriving there I had to wait for 35 minutes. The cabin crew was not making the flight comfortable either. Their manner of addressing passengers was abrupt and the service to some passengers was forgotten.

In these cases, it is clear that ground staff, individuals responsible for Internet service, the pilots and flight attendants do have room to improve the comfort score and increase the likelihood of a passenger choosing to fly again with the company. In these cases, an expensive, very comfortable seat would not have an effect on the ticket sales. I would like to stress that in those cases it is not only the designer, engineer, supplier, or manufacturer that determines the comfort rating, but management and other employees of the airline often have an even larger influence. Self-service check-in (online or at the airport) can improve comfort when it works, but has a negative effect when the system does not work. There, of course, is a limit to how much passengers should do for themselves. The moment passengers decide on the amount of fuel to load on an airplane, passenger comfort could decline dramatically.

REFERENCES

Blok, M., P. Vink, and I. Kamp. 2007. Comfortable flying: Comfort in aircraft interiors seen through the eyes of the passengers (in Dutch). *Tijdschrift voor Ergonomie* 32 (4): 4–11.

Bordass, W., and A. Leaman. 1997. Strategic issue in briefing, design, and operation future buildings and their services. Strategic considerations for designers and clients. *Building Research and Information* 25 (4): 190–195.

Brauer, K. 2004. Convenience, comfort, and costs. Presentation at the aircraft interior EXPO 2004. 30 March, Frankfurt.

Bronkhorst, R. E., L. F. M. Kuijt-Evers, R. Cremer, J.W. van Rhijn, F. Krause, M. P. de Looze, and J. Rebel. 2001. *Emotion and comfort in cabins*, (in Dutch). Report TNO, Hoofddorp: TNO Arbeid. Publ.nr. R2014871/4020054; confidential.

Bubb, R. 2008. Sitting comfort. Paper presented at IQPC aircraft interior innovation. 11 November 2008. Hamburg.

Distel, H, and R. Hudson. 2004. Judgement of odor intensity in influenced by subjects' knowledge of the odor source. *Chemical Senses* 29: 199–208.

Egmond, R. van. 2008. The experience of product sounds. In *Product experience*, eds. N. J. Schifferstein and P. Hekkert. Amsterdam: Elsevier, 69–90.

Goossens, R. H. M. 1998. Measuring factors of discomfort in office chairs. In *Global ergonomics*, ed. P. A. Scott. Proceedings of the Ergonomics Conference. Amsterdam: Elsevier Science.

Goossens, R. H. M., R. Teeuw, and C. J. Snijders. 2005. Sensitivity for pressure difference on the ischial tuberosity. *Ergonomics*, 48(7): 895–902.

Hamberg-van Reenen, H. 2008. *Physical capacity and work related musculoskeletal symptoms*. Proefschrift, Vrije Universiteit, Amsterdam.

Helander, M. G., and L. Zhang. 1997. Field studies of comfort and discomfort in sitting. *Ergonomics* 40: 895–915.

Herzberg, F., B. Mausner, and B. B. Snyderman. 1959. *The motivation to work.* Somerset, NJ: Transaction Publishers.

Lee, S. Y., and J. L. Brand. 2005. Effects of control over office workspace on perceptions of the work environment and work outcomes. *Journal of Environmental Psychology* 25: 323–333.

Looze, M. P. de, L. F. M. Kuijt-Evers, and J. H. van Dieën. 2003. Sitting comfort and discomfort and the relationships with objective measures. *Ergonomics* 46: 985–997.

Mergl, C. 2006. Entwicklung eines verfahrens zur optimierung des sitzkomforts auf automobilsitzen, PhD disser. Technical University, München.

Nefs, H. A. T. 2008. On the visual appearance of objects. In *Product experience*, eds. N. J. Schifferstein and P. Hekkert. Amsterdam: Elsevier, 9–40.

Nordin, M. 2005. Zusammenhang zwischen Sitzen und arbeitsbedingten Rückenschmerzen. In *Ergomechanics*, ed. H. J. Wilke, (10–35). Aachen, Germany: Shaker Verlag.

Parent-Thirion, A., H. F. Macías, J. Hurley, and G. Vermeylen. 2007. Fourth European Working Conditions Survey. The European Foundation for the Improvement of Living and Working Conditions, Dublin.

Rosmalen, D. van, L. Groenesteijn, S. Boess, and P. Vink. 2009. Seat comfort requirements for watching a screen. *Journal of Design Research*, 8(1): 87–100.

Ross, P. 2008. Ethics and aesthetics in intelligent product and system design. PhD thesis Technical University, Eindhoven, The Netherlands.

Schifferstein, N. J., and P. Hekkert, eds. 2007. *Product experience*. Amsterdam: Elsevier.

Sonneveld, M. 2007. Aesthetics of tactual experience. PhD thesis, Technical University, Delft, The Netherlands.

Theimer, E. T., ed. 1982. *Fragrance chemistry: The science of the sense of smell.* New York: Academic Press.

Vink, P., ed. 2005. *Comfort and design.* Boca Raton, FL: CRC Press,

Zenk, R. 2008. Objektivierung des Sitzkomforts und seine automatische Anpassung, PhD thesis, Technical University, München.

Zhang, L., M. G. Helander, and C. G. Drury. 1996. Identifying factors of comfort and discomfort in sitting. *Human Factors* 38 (3): 377–389.

2 Other Aircraft Interior Comfort Studies

Overview: The study described in Chapter 3 is not the only study on aircraft interior comfort. Some impressive studies have been done before and are presented in this chapter. Konieczny (2001) wrote a PhD thesis on this topic. He distinguishes various phases in the comfort experience and concludes that preflight experiences do influence comfort. One of the first studies was done in 1977. It is interesting to see that knee space was the major problem in that study as well as in a more recent study of Blok, Vink, and Kamp (2007). Of course, the problem mentioned by passengers in 1977 on tobacco smoke is no longer valid. A Taiwanese study stresses the importance of staff, a good Web site, and in-flight entertainment (IFE) just as other chapters of this book will show. A U.S. study stresses the importance of seat width at eye level and the triple seat configuration.

LACK OF MANY SUBSTANTIAL STUDIES ON AIRCRAFT COMFORT

Of course, it is wise to first find out what has been studied before regarding aircraft interior comfort. However, the number of studies in the scientific literature regarding aircraft interior comfort is relatively small. One explanation for this could be that there is a great deal of research in this area among manufacturers and airlines, but they have doubts about sharing this information with the public. It is certainly possible that an aircraft seat manufacturer will not share its unique comfort research with others. One can assume aircraft seat manufacturers and airlines have conducted significant research given the progress in comfort that has been demonstrated and the many innovations shown at aircraft interior trade fairs and in the magazine *Aircraft Interiors International.* Another explanation could be that when a product is doing well there is no need to research it further. On the other hand, it is important to share knowledge because we know that passengers come to expect a consistency in comfort on the airlines.

Searching for literature in the Science-Direct database on the term *aircraft interior comfort,* most of the papers that pop up concern sound/noise and air quality. On February 1, 2009, 44 percent of the papers consisted of studies of sound/noise and 29 percent of air quality. The search in the Scopus database shows a similar result. The rest of the studies mostly concentrate on very specific parts of the aircraft. Rickenbacher and Freyenmuth (2008) described, for instance, a new pneumatic system for a business seat, which has a firm upright position and a comfortably soft reclining position. There also are papers discussing the methodology on how to study comfort. Brindisi and Concilio (2008) introduce, for instance, an approach for modeling passengers' perceptions about environmental comfort inside an aircraft cabin by neural networks. In this chapter, however, more generic comfort studies will be described that are relevant to aircraft interior comfort during the flight. In the studies mentioned below, it will be clear that factors do influence one another. The preflight experience influences the in-flight comfort and an increase in noise can even make passengers experience more neck pain compared with the situation without noise. It could be that the human system works in such a way that when noise is a disturbance, one feels irritated and the resulting tension results in more pain in the neck area. In one of the studies, the large influence that staff has on passengers is shown. In the service experience, the willingness of the staff was the most important factor in the Taiwanese study, which is mentioned below.

A CLASSIC STUDY

Richards and Jacobson (1977) were one of the first to study passenger comfort. They questioned 861 passengers. Their outcomes are still interesting because some results still remain valid. They calculated the gamma coefficient for several factors influencing comfort. This is a statistical method in which, if the gamma coefficient is higher, the probability that the rank ordering of the two variables agree is also higher. The most influential factors back in the 1970s were: leg room, seat characteristics, and movements of the aircraft. Unwanted tobacco smoke from neighbours

TABLE 2.1
Gamma Coefficients between Rated Sources of
Discomfort and Overall Comfort Judgments

Factor	Comfort (%)
Leg room	.54
Seat firmness	.54
Seat width	.52
Seat shape	.51
Workspace	.49
Side motion	.48
Seat adjustment	.47
Up/down motion	.46
General vibration	.44
Sudden jolts	.43
Noise	.41
Back/forward motion	.40
Sudden descents	.35
Ventilation	.31
Turning	.28
Lighting	.27
Temperature	.27
Pressure	.26
Tobacco smoke	.23
Odors	.15

Source: Adapted from Richards, L. G., and I. D. Jacobson. 1977. *Ergonomics*, 20: 499–519.

also was mentioned by passengers. Of course, the problem of tobacco smoke is no longer encountered.

As in the current study, leg room is an important factor. Richards and Jacobson also found that there is a large increase in percentage of satisfied passengers when leg room is increased from 24 inches (61.0 cm) to 27 inches (68.6 cm). As indicated in Table 2.1, the people who state that the seats are not wide enough or that there is not sufficient leg room tend to rate their flights as less comfortable.

GERMAN STUDY ON AIRCRAFT INTERIOR COMFORT

An impressive and more recent study regarding comfort in airplanes was done by Konieczny (2001). He distinguishes hardware and software factors in studying aircraft interiors (Table 2.2). Hardware covers elements in the interior including seats, IFE, and storage systems for luggage.

Software is connected to aspects, such as meals, delays, and information. Additionally, Konieczny distinguishes lifeware (crew, neighbours, personal characteristics) and reputation. For each of these aspects, further specifications were

TABLE 2.2
Discrete Elements Influencing Aircraft Passenger Comfort

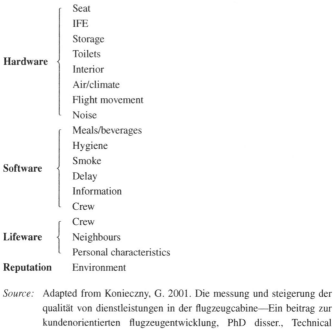

Hardware	Seat
	IFE
	Storage
	Toilets
	Interior
	Air/climate
	Flight movement
	Noise
Software	Meals/beverages
	Hygiene
	Smoke
	Delay
	Information
	Crew
Lifeware	Crew
	Neighbours
	Personal characteristics
Reputation	Environment

Source: Adapted from Konieczny, G. 2001. Die messung und steigerung der qualität von dienstleistungen in der flugzeugcabine—Ein beitrag zur kundenorientierten flugzeugentwicklung, PhD disser., Technical University, Berlin.

made. He studied the flight comfort before, during, and after the flight using 936 subjects. These subjects had to complete a questionnaire, and Konieczny then used an exploratory factor analysis to analyze the results.

It is interesting to see that the comfort model Table 1.1 in this book has some characteristics that fit very well to his study. Konieczny also distinguishes different comfort stages:

- The expectations: The personal attitude toward flying and fear for flying, but also reputation of the company
- The experience preceding the flight at the airport
- The experience during the flight
- The experience after the flight

EXPERIENCE PRECEDING THE FLIGHT

Attitude, fear, and airline reputation do influence the preflight experience, as does hardware, such as airport signs, software (e.g., waiting), and lifeware (e.g., staff competencies) (Table 2.3). Konieczny (2001) established correlation of factors with comfort preceding the flight (e.g., Figure 2.1). He found the highest correlations with fear

TABLE 2.3
Discrete Elements Preceding the Flight Influencing Aircraft Passenger Comfort

Hardware	Airport signs
	Walking distance
	Seating possibilities
	Toilets
	Shopping possibilities
	Smoking facilities
Software	Waiting
	Boarding
Lifeware	Staff competencies
	Personal support

Source: Adapted from Konieczny, G. 2001. Die messung und steigerung der qualität von dienstleistungen in der flugzeugcabine—Ein beitrag zur kundenorientierten flugzeugentwicklung, PhD disser., Technical University, Berlin.

FIGURE 2.1 The passenger view when sitting at a table in a business lounge could even have some influence on the comfort experience in the plane.

for flying (r = 0.492), attitude toward flying (r = 0.366), and airport signs (r = 0.301, Figure 2.2). The first two are difficult to influence, but a great deal of knowledge is available for improving airport signs. Generalizing this study to other countries should be done with care. These correlations could be different in other countries, as most subjects were Germans. In addition, German airports are generally well organized and it could be that, in other cultures, people have more problems with long waits.

FIGURE 2.2 Even airport signs influence the comfort experience.

EXPERIENCE DURING THE FLIGHT

Again, past experience does influence the flight experience. Konieczny found that comfort during the flight correlates most highly with the comfort preceding the flight (r = 0.407), fear of flying (r =0.492), and attitude toward flying (r = 0.367). Correlations to other factors are weaker.

For example, 55.1 percent of the variance of comfort during the flight can be explained by eight groups of factors. In fact, these are the most important factors influencing the flight comfort. These include (in order of importance, with the most important first):

- Comfort preceding the flight
- Fear of flying
- Space in the seat
- Familiarity with the aircraft
- Waiting for the end of the flight
- Attitude toward flying
- Reputation of the airline
- Backrest adjustment

The interesting part of this study is that the relative importance of the several factors helps set priorities for what should be improved first.

Many airports and airlines try to improve preflight comfort. In the perspective of this study, it is worthwhile to pay attention to the preflight experience. In one airport a short presentation is shown to tourists at the end of their visit, including an overview of memorable parts of the country they just toured in combination with their flight preparation. The service is intended for people flying on all outbound flights of a specific airline and includes ticket and passport control, security checks, luggage weighing, delivering boarding passes, luggage check-in, and transporting the luggage to the airport. This significantly reduces the amount of time required of tourists to arrive at the airport before the flight, and they are able to reach passport control only 55 minutes before takeoff. The time saved can be used to tour the presentation site, concluding the visit to the country. This is an extreme form of influencing the preflight comfort experience, but it does have its influence and can attract passengers.

EXPERIENCE AFTER THE FLIGHT

This experience was mainly dominated by the expectations and the experience preceding and during the flight. The correlations between comfort after the landing and other factors were highest for comfort preceding the flight ($r = 0.407$) and during the flight ($r = 0.563$). Other factors of some importance included travel distance at the airport ($r = 0.19$) and airport signs ($r = 0.189$).

This excellent German study provides valuable information and shows that the preflight and the flight itself are major factors influencing the total comfort, which is strongly related to "flying again" with the same company and, thus, considered worthwhile.

A STUDY REGARDING SERVICE, PERCEIVED VALUE, AND SATISFACTION IN TAIWAN

A study done in a country with a culture different from the Western culture is the Taiwanese study mentioned in the Overview. This interesting study was done with 300 passengers flying international routes (Chen, 2008).

Like Konieczny (2001), Chen used an exploratory factor analysis to analyse the results. He also added a principal component analysis, which made it possible to identify the factors explaining the service quality in airplanes (Table 2.4). Simply put, Chen found factors of importance for "service." The most important factor was the staff and facilities. It explained 19 percent of the variances. Within that factor, helpfulness was the factor having the most influence. Another important factor was product. Because this study was focused on service and not on comfort, the hardware parts of the airplane were not mentioned in the questionnaire. Regarding service, communication was seen as important (e-mail, Internet, etc.) as well as up-to-date IFE.

The study shows that apart from the willingness of staff it is important to have the Web site running well. It should have sufficient information and a good booking function.

A GERMAN STUDY OF NOISE

Another interesting study on aircraft interiors was Mellert et al. (2008), which considered noise. They studied the impact of noise and vibration on well-being during long-haul flights as well as in aircraft simulators. Apart from indices to characterize the human response, they found that noise has an important impact on health indicators, comfort, and well-being. For instance, passengers with swollen feet are more aware of this situation under noisy conditions. The awareness increased 43 percent under noisy conditions compared with the quiet conditions in the beginning of the flight. The same is true for neck pain. A pronounced increase in pain occurred with increasing noise levels, according to the study. Another interesting finding in this study is that the perception of air quality reduces during the flight. The degradation

TABLE 2.4

Results of the Exploratory Factor Analysis of Service Expectation

	Factor Loading	Variance Explained (%)
Factor 1: Employees/facilities		19%
Willingness to help from staff	0.72	
Courtesy of staff	0.7	
Prompt and correct service	0.65	
Cleanliness of staff	0.64	
Baggage loss and damage handling	0.64	
Efficient booking queuing line	0.63	
Sufficient checking in and baggage handling service	0.61	
Interior cleanliness	0.59	
Good safety image of airline	0.57	
Factor 2: Product		17%
Internet, e-mail, fax, and telecom service on flight	0.73	
Up-to-date entertainment on flight	0.71	
Frequent flier program	0.65	
Prompt food and beverage service	0.6	
Sufficient food and beverage on flight	0.59	
Provision of preferred seat option	0.57	
Up-to-date aircraft and facilities	0.55	
Global air alliance service	0.55	
Tax-free commodities	0.52	
Individual care from staff	0.5	
Factor 3: Transaction		13%
Sufficient information on Web site	0.75	
Booking function on Web site	0.75	
Correct reservation service	0.69	
Provision of flight information	0.61	
Convenient reservation service	0.59	
Factor 4: Reliability	·	12%
Doing things right the first time	0.76	
Punctuality	0.73	
Convenient schedule	0.72	
Confidence in the staff	0.56	

Source: Adapted from Chen, F. C. 2008. Investigating structural relationships between service quality, perceived value, satisfaction, and behavioural intentions for air passengers: Evidence from Taiwan. Transportation Research Part A 42, pp. 709–717.

of the perceived air quality was 15 percent, while there was no indication of an objective change of air quality.

This shows that questioning passengers about noise is difficult because the noise itself may not be mentioned by passengers as a problem, but it does influence other problems. The same could be true for lighting and cabin air quality. These two items could influence complaints on other issues on which passengers report.

A DUTCH STUDY REGARDING AIRCRAFT INTERIOR COMFORT

Blok, Vink, and Kamp (2007) studied 291 trip reports of passengers (Figure 2.3). They mentioned that a shortcoming of trip reports is the fact that the elderly may not commonly use the Internet, and it could be that people who have complaints use Internet trip reports more freely to vent their feelings. As a result, interviews were added to the study.

In the study, 152 subjects were interviewed just after their flight and asked on a questionnaire about their comfort experience at different stages—from check-in to flying. These questions were asked of passengers directly following their flight after

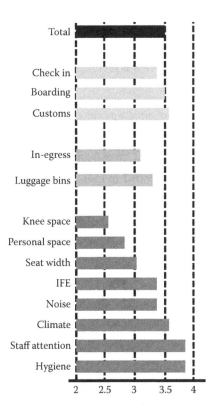

FIGURE 2.3 The comfort score on a scale from 1 to 5 (1 = no comfort; 3 = average; 5 = very good comfort) on different aspects before the flight (light grey), getting in the seat (darker grey) and while seated (darkest grey), according to a study of Blok, Vink, and Kamp (2007).

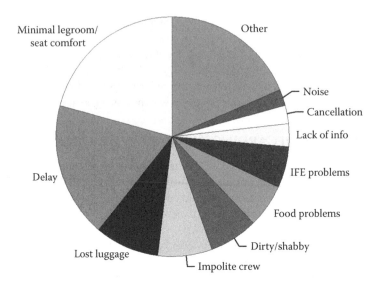

FIGURE 2.4   The problems described in the 291 trip reports analyzed in the study of Blok, Vink, and Kamp (2007).

passing customs. The questionnaire was pretested using 14 subjects and then adapted (Figure 2.4). For each part of the flight, a score could be given from 1 to 5 (1 = very bad, 2 = bad, 3 = average, 4 = good 5 = very good).

The mean of all 152 questionnaires was calculated and some tests (t-test, p < .05) were performed on specific questions. Examples of such questions included: Do taller persons differ in comfort score from shorter ones and do long-haul flights have a different comfort score from short flights? The passengers flew with 36 airlines and had an average height of 176 cm (5 ft. 7 in.).

The results are shown in Figure 2.3. It was surprising to see that none of the aspects of the mean comfort level was rated good (4) or very good (5). Knee space was rated lowest followed by personal space and seat width. In fact, the old study from 1977 showed the same result. The study showed also that taller persons need special attention. The lack of knee space is seen as a problem for taller subjects. Statistical analyses shows that taller subjects (>173 cm [5 ft. 6 in.]) rate the comfort significantly lower (t-test, t = 1.98; p < .05). The study showed that, especially for longer flights (over five hours), IFE and staff attention become more important as these significantly influence the total comfort.

During these longer flights, the comfort score for these aspects was significantly lower than during short flights. There also was an open question: What aspect needs to be improved first? Forty-one percent answered: Leg space.

As was mentioned earlier, 291 trip reports were reviewed in the same study (see Figure 2.4). The main problems are again leg room and seat comfort, followed by delays and lost luggage. That doesn't mean that almost 1 out of 12 flights is delayed. Because a passenger has to take time to make a trip report, it might be that relatively more people who have problems complete these forms.

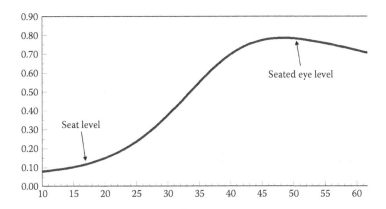

FIGURE 2.5 The correlation (y axis) between cabin width per seat and passenger preference (r^2) at different heights of measurement (x axis) above the floor in inches (Brauer, 2010).

It is also interesting in this study that the newer aircraft are rated significantly better regarding comfort than older aircraft.

A U.S. STUDY REGARDING PASSENGER EXPERIENCE

Brauer (2010) studied preference and comfort data for intercontinental flights from various sources and related these to the attributes of aircraft interiors. A comparison was made between passengers' stated preferences for the airplane they were flying in and the cabin width per seat at different heights above the floor in that airplane. This comparison showed that the correlation between cabin width per seat and passenger preference was highest at seated eye level (Figure 2.5). Another interesting finding was that having no neighbour was a large influence on passenger comfort. Because demand for air travel varies so greatly by time of day, day of week, and season, there remain empty seats on many flights in profit-maximizing airline systems. In triple seats (three-seat units, e.g., in a 3-3-3 configuration), each of these empty seats makes two passengers more comfortable. In doubles (two-seat units) and in practice in quads (four-seat units), each empty seat makes only one passenger more comfortable. Combining these two facts has substantial effects. Adjacent empty seats and width per seat at eye level explained more than 92 percent of the variation in passenger preference for airplanes (R^2).

SOME CONCLUSIONS

Several studies indicate that increasing leg room, knee space, and personal space have a positive effect on the comfort experience. So, leg room and personal space have priority in the design. However, all studies show clearly as well that not only physical aspects play an important role. Expectations and preflight experiences also need attention. When you are designing a new interior or are in the process of purchasing one, it's hard to imagine that the airport signs influence the comfort score recorded by passengers, but at least one study shows there is a relationship. The fact that the reputation of the airline, its Web site, and brochures have influence is a bit easier to understand. From psychological research, we know that our behaviour is unconsciously influenced by priming, which is the process whereby an earlier stimulus influences response to a later stimulus. Sometimes priming people with only a few words can make a difference in their behaviour. Bargh, Chen, and Burrows (1996) showed that people who were covertly primed with words relating to old age walked much slower after the experiment than people who were primed with nonage-specific words. When the priming is positive, the brain's automatic activation can have a similarly significant effect on subsequent behaviour. For instance, studies have shown that people primed with words related to "success" subsequently perform much better on intelligence tasks. A similar phenomenon might occur here if the passenger expects the interior to be comfortable and their brain is thinking "comfort," then the interior might be experienced as more comfortable.

REFERENCES

Bargh, J., M. Chen, and L. Burrows. 1996. Automaticity of social behavior: Direct effects of trait construct and stereotype activation on action. *Journal of Personality and Social Psychology* 71 (2): 230–244.

Blok, M., P. Vink, and I. Kamp. 2007. Comfortable vliegen: Comfort van het vliegtuiginterieur door de ogen van de gebruiker. *Tijdschrift voor Ergonomie* 32 (4): 4–11.

Brauer, K. 2010. Redesigning the passenger experience. Paper presented at Stanford Graduate School of Business, Stanford, CA, April 20.

Brindisi, A., and A. Concilio. 2008. Passengers' comfort modeling inside aircraft. *Journal of aircraft* 45 (6): 2001–2009.

Chen, F. C. 2008. Investigating structural relationships between service quality, perceived value, satisfaction, and behavioural intentions for air passengers: Evidence from Taiwan. Transportation Research Part A 42, pp. 709–717.

Konieczny, G. 2001. Die messung und steigerung der qualität von dienstleistungen in der flugzeugcabine—Ein beitrag zur kundenorientierten flugzeugentwicklung, PhD disser., Technical University, Berlin.

Mellert, V., I. Baumann, N. Freese, and R. Weber. 2008. Impact of sound and vibration on health, travel comfort, and performance of flight attendants and pilots. *Aerospace Science and Technology* 12: 18–25.

Richards, L. G., and I. D. Jacobson. 1977. Ride quality assessment III: Questionnaire results of a second flight programme. *Ergonomics*, 20: 499–519.

Rickenbacher, U., and K. Freyenmuth. 2008. Lantal promises pneumatic comfort for airline passengers. *Advances in Textiles Technology*, June, 6–7.

3 The Voices of over 10,000 Customers

Overview: In this chapter, 10,032 trip reports of passengers flying in 2009 are analysed. The airline industry performed well with newer planes receiving a higher average comfort rating than older ones (7.75 on a scale from 0 to 10; older ones scored 6.2). The factors having a large influence on comfort were leg room, hygiene, and crew, but the influence of the seat is also substantial. Delays and lost luggage occurred in only a minority of cases, but in those cases the comfort rating is dramatically lower. Expectations also are of importance and the attention of the staff is a great opportunity to increase the comfort experience because it can be realized with relatively low investments. This would include familiarizing the staff with the results of this study and providing relevant training. Cleanliness is highly related to comfort as well. The aspects mentioned by the passengers in this study provide considerable input to redesign opportunities.

TECHNOLOGY VERSUS PASSENGER

In the previous chapters, a survey among passengers was often mentioned. In this chapter, the results of this study among 10,032 are described in more detail. The study does not focus on what is technologically feasible or interesting to do, but it focuses on what passengers currently experience. This experience is important because they are customers making new purchase decisions each day. According to some authors, the need to economize (Pine and Gilmore, 1999) is getting more important. At home, people pay less than 20 eurocents on a cup of coffee, but, for an extra experience, people are willing to pay up to 25 times more. At the St. Mark's Square in Venice, a cup of coffee costing five euros is not unusual. The example shows the importance of additional experiences, which can be applied readily in the current airline industry. Recently, considerable knowledge has become available in the field of product experience (Schifferstein and Hekkert, 2008). One clear fact is that listening to the voice of the customer is critical. The customer's voice is the central theme of this chapter. This voice may be another point of view from the normal, but it does reveal what passengers experience, and the airline industry can use it as a source of inspiration. The experience of passengers has changed over the past decade. Nowadays, travelling consumes a significant amount of our time, and, with increasing frequency, the passenger of today is being presented with new developments regarding Internet check-in, lounges, new seats, in-flight entertainment (IFE), meals, flat beds, lighting, and other interior changes. The question is whether these new innovations are noticed by passengers and what do passengers like or prefer.

Whatever the motivation for travel, humans are the user and a great many spend a significant amount of their time travelling, which is an important reason to travel in the most efficient manner and in the most pleasant surroundings possible. This was observed a few decades ago by Oborne (1978). For an airline, it is important to adapt the transport environment to the passenger and give the passenger the experience, which can be better than expected. If the passenger becomes very dissatisfied with his journey, he is likely to take his business elsewhere. Comfort is an opportunity for a unique selling point. However, as was described before, when more companies increase their comfort level, it is more difficult to differentiate them in regard to comfort.

There is much going on in the field of the aircraft interior. For instance, at the aircraft interior EXPO 2010 in Hamburg, many innovations were shown by over 500 exhibitors including Airbus and Boeing. Over 10,000 visitors attended this EXPO in 2010.

INNOVATIONS HAVE THEIR EFFECTS

The previous chapters show clearly that manufacturers are active in innovating. A study in 2007 (Blok, Vink, and Kamp, 2007) showed that the newer aircraft earn a significantly higher average comfort rating than older aircraft. In this case, the older aircraft include the A300, A310, Boeing 737-300, and 737-400, and the newer aircraft: the A330 and Boeing 737 Next Generation. The same effect was found in

TABLE 3.1
The Experienced Comfort of Older Planes and Newer Planes

	Mean Comfort	STDEV	N
Older planes	6.22	2.49	667
Newer planes	7.51	1.83	1,472

the present survey. Not all survey respondents mentioned the aircraft type precisely or they mentioned just Airbus or Boeing 737. This is not enough information to make a separation in the age of the plane. But 2,139 passengers did explain the type well enough to make a distinction in the plane's age. For instance, the Boeing 737-200, 300, 400, and 500 series can be seen as the older types and the 600, 700, 800, and 900 as the newer types. Boeing defines the latter as the next generation 737. The A300, A310, 757, and 767-200 are seen here as the older planes and the newer planes are the A330, 777, and 767-400.

The difference in comfort experience is shown in Table 3.1. The difference is highly significant ($p < 0.0001$, two-sided t-test) as in other studies. So, newer airplanes are found to be more comfortable.

STUDY METHODOLOGY

Before describing the rest of the results, it is important to describe the survey methodology. The study was initiated by a group of airlines, which also partici-pated in the study. During flights of these airlines, passengers were informed that there is a Web site available where they could make a trip report. Awards, like free hotel nights, were raffled among the respondents by an independent party. The trip reports could be written in an open text box; photos also could be attached. At the end, some questions were asked regarding the comfort experience (a number from 0 to 10, 10 = maximum comfort), the airline flown, if they would book a flight again for a trip with this airline (yes, no, don't know), and what class they were in (economy/coach, premium economy, business, or first). The 10,032 passengers who completed the trip reports travelled with 123 airlines. Most of these airlines were from Europe (40.9%, Figure 3.1) and North America (31%). There were 162 photos were uploaded and used to illustrate the problems or positive aspects. Also, 69 passengers gave us explicit permission to use the photos in public reports (see Chapter 6).

Nine students analyzed the trip reports and input this data into files that were later merged into one large file. For each trip report, flight characteristics were input into the file. Characteristics included which airline was used, the flight duration, comfort score, whether or not it was a direct flight, the knee space, the seats, any delays, lost luggage, the crew attitude, hygiene of the plane, and further remarks.

Statistical analyses consisted of correlation calculations and, as comfort was the main issue being tested (t-test, $p < 0.5$), whether a good score on one characteristic had a significant impact on comfort compared with a bad score. For instance, the

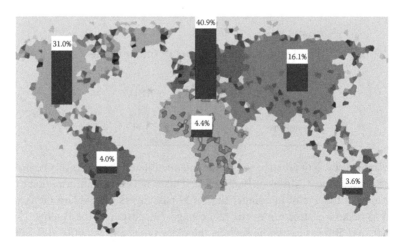

FIGURE 3.1 Domicile of the airlines evaluated by passengers in this survey.

comfort score of the individuals mentioning a positive crew attitude was compared with the group mentioning a negative attitude, and the impact on the comfort score was established. More extensive statistics were used in a scientific research paper by Vink et al. (2011), but not included in this book.

In this chapter, the correlations will be described and then each aspect influencing comfort will be discussed and illustrated with quotes of the passengers. Some photos of passengers will be shown that illustrate the main findings. (For more photos on the subject, see Chapter 6.)

Factors Correlating with Comfort

The factors most often mentioned in the trip reports included: nice or rude crew (which could be ground staff, pilots, or flight attendants), followed by leg room or pitch, delay, lost luggage, seat aspects, IFE, hygiene, and direct flight. These factors correlate with comfort (Figure 3.2) and are of somewhat more importance because they are mentioned by more than 8 percent of the passengers. The fact that they are mentioned is noteworthy as it shows that it is an issue that many passengers notice and have on their minds, meaning they are conscious of these issues.

Other factors, such as climate, air quality, noise, toilet, waiting, information, food, neighbour, and luggage room are mentioned in less than 8 percent of the cases, but still correlate with comfort (Figure 3.3). This suggests that passengers are not always aware of these factors. On the other hand, we know from other studies that these factors influence the experienced factors. In Chapter 2, a study is mentioned in which passengers with swollen feet were more aware of this situation under noisy conditions.

The factors that correlate highest with aircraft interior comfort are "fly again," leg room, hygiene, and the crew.

"Fly again" means that if the comfort is high there is also a good chance that a passenger would book a flight again for this trip with this specific airline. This is in alignment with other findings that an airline can attract passengers by increasing

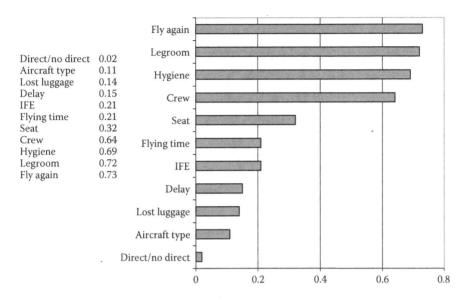

Direct/no direct 0.02
Aircraft type 0.11
Lost luggage 0.14
Delay 0.15
IFE 0.21
Flying time 0.21
Seat 0.32
Crew 0.64
Hygiene 0.69
Legroom 0.72
Fly again 0.73

FIGURE 3.2 Correlations between comfort score and aspects mentioned in more than 8 percent of the trip reports. A higher correlation means a stronger relationship between comfort and the factor.

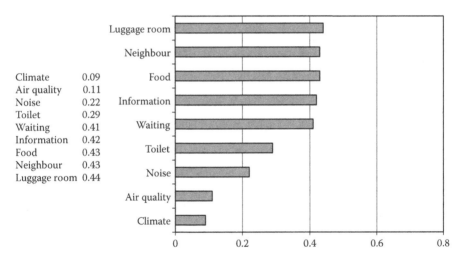

Climate 0.09
Air quality 0.11
Noise 0.22
Toilet 0.29
Waiting 0.41
Information 0.42
Food 0.43
Neighbour 0.43
Luggage room 0.44

FIGURE 3.3 Correlations between comfort score and aspects mentioned in less than 8 percent of the trip reports.

the comfort (Brauer, 2004). The relationship between comfort and leg room is also described in other studies (Blok, Vink, and Kamp, 2007; Richards and Jacobson, 1977), and it is the most important factor related to comfort. Less leg room results in lower comfort ratings. Some passengers and even persons working in the airline industry relate leg room directly to pitch (the fore–aft distance between like points on seats in adjacent rows, e.g., from the back of one seat to the back of the seat in

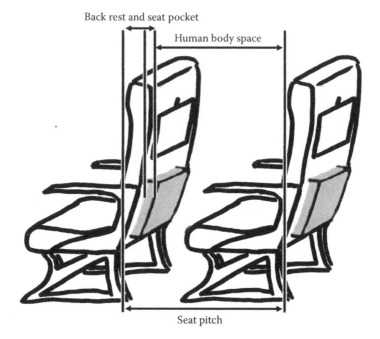

FIGURE 3.4   A figure showing that pitch alone does not determine the leg room.

the next row). However, we should be careful in relating leg room directly to the pitch. From the answers of the passengers in the trip reports, some passengers even choose their plane based on pitch. The next seat row can be, for example, positioned 34 inches forward. However, if you have a thick back rest and a full seat pocket in front of your knees, this reduces knee room (Figure 3.4). It often happens that a seat with a thinner back rest at a lower pitch gives you more knee room.

The correlation between hygiene and the comfort score is often high when a clean or fresh looking airplane is mentioned. It seems worthwhile to have clean-looking planes, which means paying attention to cleaning and to buying products that retain their looks for a long time. Also, in this study, it is shown that the crew is crucial in giving passengers a comfortable flight. The crew paying positive attention and giving clear messages is crucial for passengers to experience a comfortable flight.

For the issues mentioned in less than 8 percent of the trip reports, there is a correlation between a particular issue and comfort, which means that there is a relationship between a lower comfort and:

- Not enough hand luggage room
- A terrible neighbour, or a seat that makes if difficult to get pass a neighbour
- Lack of food or bad food
- Not enough information, too often loud information on the intercom, or a message that is very hard to understand
- Waiting before boarding and deboarding, and waiting for other services

FIGURE 3.5   This tall passenger could still fly because there is no neighbour.

In the following paragraphs, the aspects mentioned in Figure 3.2, as well as luggage room and neighbour, will be elaborated on to explain passenger likes and dislikes.

LEG ROOM

Leg room is strongly connected to aircraft interior comfort. This is shown by the correlation coefficient, which is 0.72 between leg room and comfort score. Twenty-seven percent of the respondents mentioned pitch, knee space, or leg room in the trip reports. If we compare the passengers giving a positive comment on the leg room (11 percent) with the passengers giving negative comments (16 percent), the difference between the comfort scores of these groups is highly significant (two-sided t-test, $p < 0.0001$, $t = 13{,}57$). This means that such differences are extremely unlikely to occur by chance.

Following are some examples of positive comments made by the passengers in the reports. In charter flights, a pitch of 33 inches is called "comfortable," as passengers probably expected less space. A smart position of the seat pocket for the magazine at headrest level is seen as positive and a special seat, such as a seat next to the exit row, having no neighbours (Figure 3.5), or receiving an upgrade to another class, of course, is seen as very positive because of the extra leg room. Sometimes taller passengers do report that they are astonished that they fit in the seat as they normally have difficulty. The possibility of being able to stretch ones legs under the seat in front and change position is also mentioned several times.

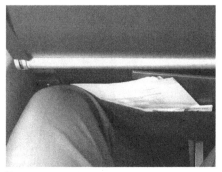

FIGURE 3.6   Two photos of inadequate leg room. The image on the right shows clearly how the seat pocket reduces knee room.

There also were some negative comments. Often, we found the opinion that it was fortunate that the flight was not longer because the knees would not have held out much longer. Sometimes people mentioned the cramped or tight seating positions. The reason why the seat pocket is placed at the position where you need the most space is also discussed in the trip reports (Figure 3.6). It was interesting to see that the number of complaints was around 10 percent both in short- and in long-haul flights, while we expected more complaints in the long-haul flights because knee space might be a bigger problem when one is sitting longer in this position. On the other hand, it could be that long-haul planes have somewhat more leg room.

HYGIENE

Hygiene also is strongly connected to aircraft interior comfort. Twelve percent of the passengers mentioned something in their report on an aspect connected to hygiene. It could be a clean plane or dirt in the windows. The correlation coefficient with comfort is .69. If we compare the passengers giving a positive comment regarding hygiene with the passengers giving a negative comment, the difference between the comfort scores of these groups is highly significant (two-sided t-test, $p < 0.0002$, $t = 3.95$). The average comfort score of those complaining about hygiene was 4, and the score of the passengers giving positive comments was 7.9.

Examples of positive hygiene include a clean-looking plane, a fresh interior, a proper looking interior, clean seats, a new company journal, clean windows, and a bright refurbishment.

Examples of negative hygiene seen in the trip reports include dirty cabins; old and shabby cabins; dirty, worn out cabins; dirty blankets; seats old and worn; smelly; dirt on the chair; dirt in the seat pocket (even mouldy bread and an apple core in the seat pocket were found); and dust in the window (Figure 3.7).

CREW

As stated in most of the other studies, the crew is of great importance to the comfort rating (Figure 3.8). The correlation between crew and comfort was 0.64. Most

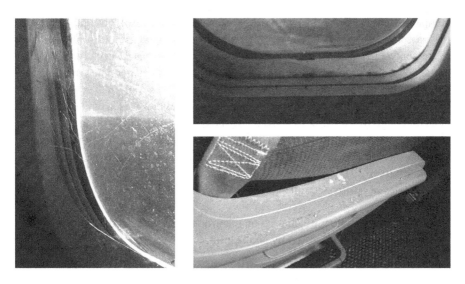

FIGURE 3.7 Pictures of travellers related to hygiene. (Left) spider web in the window, (upper right) dust on the seat, (lower right) dust in the window.

FIGURE 3.8 A warm welcome in the aircraft positively influences the comfort experience.

companies are doing well regarding this aspect. In all trip reports, 60 percent of the passengers mentioned a nice crew, and crew complaints were only found in 17 percent of the cases. This 17 percent is important because a negative score dramatically reduces the total comfort score of a flight. This is not a new finding. It is described in the study by Blok, Vink, and Kamp (2007) as well. In the current study, the average score of the flights with positive crew comments was 8, while, in the case of a complaint, this was 3.9. If we compare the passengers giving a positive comment regarding the crew with the passengers giving a negative comment, the difference between the comfort scores of these groups is highly significant (two-sided t-test, $p < 0,0001$, $t = 12,7$).

Examples of positive comments on the crew included:

- The crew was very welcoming on the ground as well as in the air.
- The cabin crew was friendly and helpful.
- Young enthusiastic crew.
- Cabin crew was very efficient.
- The flight attendants were polite and accommodating.
- The pilots gave clear and good information also about the sudden movements of the plane.
- The flight attendants were pleasant on all of the flights.
- Service from the flight attendants was excellent on the way back.
- The lounge staff even helped me find the gate.
- The ground staff arranged for the plane [to] wait a few minutes for us to arrive.
- I will never forget the smiling staff.

Examples of negative comments on the crew included:

- The ground staff was unhelpful and had personal conversations while I was in a hurry catching my plane.
- The flight crew frequently abused [sic] in their mother language to each other about the passengers.
- The flight attendant was too busy laughing and carrying on to give instructions over the loudspeaker instead of helping.
- She was rude to me, shoving potato chips in my face.
- The FA was unprofessional, there was one employee loudly routing passengers to queues.
- The pilots didn't seem to bother if their messages came through; how difficult is it to speak good English?
- The staff was extremely arrogant towards the passengers.
- The crew was reprimanding people when asked questions.
- The cabin crew did not create a very favourable impression by their demeanour.
- The crew addressed passengers abrupt and curt.
- The crew was really not interested in doing their job well.

This survey clearly shows that there is room for improvement regarding the crew. Especially, in the current experience economy where work is a theatre and every business a stage (Pine and Gilmore, 1999) it is of great importance to know some of the basic rules of entertaining. It was a surprise to see that in 3 percent of all cases the passengers do not understand the messages of pilots and flight attendants. Flight attendants mumbling "chez or chee" meaning do you want a sandwich with chicken or cheese, pilots speaking so loud that distortion is covering the message or so fast that even native English speakers cannot follow it. It's all daily practice in the air, meaning that much more money can be made here. Just make some recordings during a flight and let the crew hear the messages during a refresher course and there should be a quick turnaround. Crew attention seems a simple way to attract passengers, but considerable opportunity still remains. Improving the attention of the crew can be achieved by just showing the abovementioned examples. It is wise to remind members of the staff of their importance by informing them of the results of this study and providing relevant training.

Luggage Room

Luggage room problems are mentioned by less than 2 percent of the passengers. However, it is correlated to comfort (correlation = 0.44). The positive comments concern ease of reach (height as well as horizontal distance when standing in the aisle), enough space (Figure 3.9), and nice to have it so close and easily reachable. The latter were comments from business class passengers having storage possibilities alongside their chair. Negative comments reported included: difficult to reach, not enough space, a window seat is terrible if you want to reach your luggage. Why is there no check on the number and size of hand luggage when getting into the airplane? The passengers that board the plane last have to store the hand luggage below the seat in front of them, which drastically reduces leg space.

Neighbour

The factor "neighbour" can only be influenced to a certain extent. In business or first class, this complaint is not often seen. In general, having no neighbour or only one

FIGURE 3.9   The luggage room of the Embraer 170/195 series was praised by passengers. This also has to do with expectations because in other regional aircraft this space is often less.

FIGURE 3.10 Some airplanes have two seats abreast, which has a positive effect on the comfort experience.

results in high comfort scores (see Figure 3.4). This is not new; it also was described in Chapter 2 in the U.S. study of Brauer (2010). Some airplanes having two seats abreast (Figure 3.10) have a higher score in this area since there is only one neighbour and, while checking in, the ground staff should pay attention to the fact that, when the plane is not fully booked, there is an opportunity to leave an empty seat between two persons. In some reports, when a plane is not fully booked, airlines seem to have the habit of keeping the first and last rows unoccupied. Perhaps it is better in those cases to spread the passengers more evenly throughout the plane, of course, taking into account the infrequent cases in which aircraft balance is critical. It is difficult to change the 1 percent of passengers who do have specific complaints of their neighbour, such as, the neighbour smells bad, is dirty and shabby, is too fat, is always fighting to put his arm on the armrest, touches me too frequently, wants to pass me constantly during my sleep, is making too much noise, is snoring, he keeps laughing and talking. On the other hand, using a triple seat conversion instead of the quad or double, each empty seat makes two passengers more comfortable. And, if areas are defined in the plane for ones who want to rest and for ones who want to chat, this would solve some problems. The correlation between this factor and comfort is 0.37.

SEAT

It was noted that 19 percent of all passengers expressed their opinion on the seat in the trip reports, showing that, when speaking of comfort, the seat is an important part of a comfortable flight. The correlation between seat and comfort was not too high ($r = 0.32$), but comparing a good and a bad seat results in highly significant differences (two-sided t-test, $p < 0.0001$, $t = 11.5$). The 14 percent positive comments were

TABLE 3.2
Some Aspects of Seats Mentioned More Than 10 Times

Positive Seat Comments Mentioned More Than 10 Times

Leather seats: very comfortable

Roomy/spacious seats

The seats seem wider than usual

The seats at first look very narrow, but they are comfortable

Very nice covering (mostly a light colour is mentioned)

Adjustable headrests

Good lumbar support

Nice in-seat foot rest

Fairly new seats

Nice rounded back to put feet under the seat

Nice wings at lumbar level

Nice short rounded armrest

Good to have space under the armrest

Great that the front of the seat can move downwards

Negative Seat Comments Mentioned More Than 10 Times

Too much lumbar support

Uniforms of interior clashes with the seats colour

The seat in front of me was broken: always backwards

Pictures of comfortable leather seats are only in the magazine

The seats were dirty and gave a poor impression

The seats in the economy class appear to be small and squished

Almost no seat cushions

The plastic seats were not comfortable and sweaty

No seat pockets

Tight/cramped seats

Seats do not recline

Very little padding or lumbar support

The controls could not be removed from the seat arm

Adjusting the seat was not user friendly

Seats were terribly close to each other

Seats were uncomfortable, but cabin crew made up for this

Seat in front of me was dirty

My tray table was broken

No side headrests

My seat was broken and reclined automatically

mostly that the seats are comfortable and the 5 percent negative comments mostly mention that the seat is not comfortable, which does not afford much information. The more specific comments are shown in Table 3.2.

For business class, a common comment was that passengers were "fighting to prevent gliding out of my chair while sleeping." Even in some so-called "flat bed" situations this was the case. The passengers report that "probably my bed is flat,

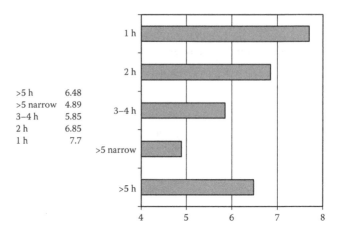

>5 h	6.48
>5 narrow	4.89
3–4 h	5.85
2 h	6.85
1 h	7.7

FIGURE 3.11   Comfort score for the hours (h) of flying on a scale from 0 to 10 (x-axis).

but the angle of the bed in space is not horizontal." Another interesting comment was that leather looks more luxurious and comfortable, but for long-haul trips, other material is preferred.

FLYING TIME

From studies of automobile driving, we see that comfort declines with additional hours of driving (e.g., Zenk, 2008). We would expect a similar consequence when flying. The longer you fly, the lower the comfort scores are. In Figure 3.11, we see that the comfort declined during four hours of flying, which is as expected. However, flying more than five hours, the average comfort of the passengers increased again. The linear correlation between flying time and comfort is low ($r = .21$), but the quadratic is much higher ($r = .71$). A possible explanation for this phenomenon could be the often-seen comment that the seats and environment in long-haul airplanes are much more comfortable as these are often wide body jets. In business class especially, the comfort in long-haul flying times is rated higher than in the business class of flights shorter than four hours. If we replace the average comfort score of all subjects by only the subjects travelling more than five hours in a narrow body jet, the comfort score is 4.89, which supports the explanation that wide body jets are more suitable for long travels. This shows again that the industry did a good job in making even longer flights comfortable, which is not easy since just remaining in place for more than five hours is in itself a source of discomfort.

IN-FLIGHT ENTERTAINMENT

The in-flight entertainment (IFE) is mentioned by 8.1 percent of the passengers. Of these, 7 percent are positive comments, usually "good IFE." IFE is not highly correlated to comfort (correlation = 0.21). Comparing a good and a bad IFE results in significant differences (two-sided t-test, $p = 0.032$, $t = 2.1$). Also, regarding IFE, there are indications that the airline industry did make some progress. In this survey 1.1 percent of the passengers had complaints about the IFE, while three years earlier in the study of Blok, Vink, and Kamp (2007), which used a similar method to analyze trip reports, 1.6 percent of the passengers had complaints (n = 291). The more specific positive comments in this survey concern a large screen, a screen that can be rotated to have the screen perpendicular to the viewing angle, a good choice of movies, that there is good choice in IFE (music, games, news, movies), good quality of sound, and active noise cancelling. The negative comments often concerned the user friendliness of the system, especially regarding the complexity of the controls. Even young passengers complain about this. Other complaints were about the bad quality of the screen and the fact that it is unclear when the movies start playing. Some passengers mentioned an IFE system from the 1970s, paying for a terrible earphone, hating the screens that are dependent on the position of the chair of the passenger in the row before you, reclining means a bad view, and having to pay so much for movies.

The problem with IFE is that the development of new systems in constantly going on and at home we probably are used to much better systems making it more difficult to meet the requirements of all passengers.

DELAY

The correlation between delay and comfort is low (0.15), but in graphics it is clear why the correlation is low. In fact, the comfort experiences decrease with a longer delay (Figure 3.12), and, for delays more than four hours, the comfort experience rises again. In a substantial part of the cases, large delays were solved in a nice way for passengers resulting in a high comfort score. This could mean a good hotel, an alternative flight, good information on the weather conditions explaining the delay, and that the company is willing to do the best it can. Nevertheless, in case of delay of more than four hours, the mean comfort score is still below average (which is 5.8) because there also are situations not solved and leading to dissatisfied passengers

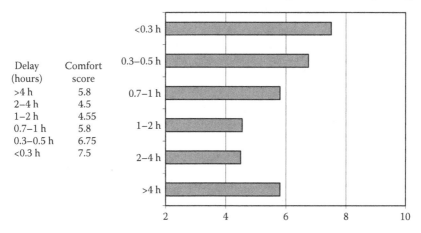

Delay (hours)	Comfort score
>4 h	5.8
2–4 h	4.5
1–2 h	4.55
0.7–1 h	5.8
0.3–0.5 h	6.75
<0.3 h	7.5

FIGURE 3.12   The comfort score (0–10) for different delay times (h = hour).

who record extremely low comfort scores. The standard deviation is high for flights of more than four hours, meaning that the variation in scores is large.

The scores of 0 were seen for unsolved problems with a lack of explanation and scores of 10 were seen when the companies solved the problems satisfactorily. This means that you can make the best interior ever, but if an airline has many unsolved delays it will not increase its comfort score. It also was interesting to see that for long-haul flights a delay of one hour was not seen as a big problem, but for short flights it was.

Lost Luggage

The correlation between lost luggage and comfort is low (0.14). This is because of large differences. Only 0.04 percent of all checked baggage was completely lost, according to the reports. Another 1.6 percent reported lost or delayed luggage. A combination of a good, comfortable flight and delayed luggage, which is brought to the hotel or home, still can have good comfort scores, especially when the information on their luggage is given to the passengers. A trip report illustrating this: "... due to the delay, the transfer time was short. At the gate of our connecting flight, the attendee told us that the bags will come later and will be brought to our hotel. That is what we call service!!...". On the other hand, lost luggage and no information could even result in comfort scores of 0. These scores were seen for unsolved problems with a lack of explanation and scores of 10 were seen when the companies solved the problems satisfactorily. This means that also, if luggage is lost, the best interiors will not compensate for this problem. This is a finding that was already described in the study of Blok, Vink, and Kamp (2007).

Aircraft Type

The correlation between aircraft type and comfort is low (0.11). This is because of the clustering into four categories: smaller, middle sized, larger, and wide bodies. If we look more into details of the type of aircraft, there are some significant differences (Figure 3.13).

We clustered the types in groups. For instance, all A300 and A310 versions are in one group, the 737-200, 737-300, 737-400, and 737-500 are forming the old 737 group, and the A318, A319, A320, and A321 are grouped into the A320 group. The A310/300 and 757 group have a significant lower score than the 767, which has a significantly lower comfort score than the others. The value of this difference is probably limited. Also, the aircraft interior even can be very different between two airplanes of the same type. We should be careful with drawing conclusions as only 2,139 passengers did specify the type precisely, and it is very difficult to interpret these data because the comfort is largely influenced by the configuration. For instance, the Boeing 777, which has three seats at the window and four in the middle (3-4-3) scored lower than the same Boeing 777 that has a 3-3-3 configuration. We can imagine that adding one seat in the same fuselage narrows the space in the seat and could reduce the comfort. This clustering lowers the average comfort score of the

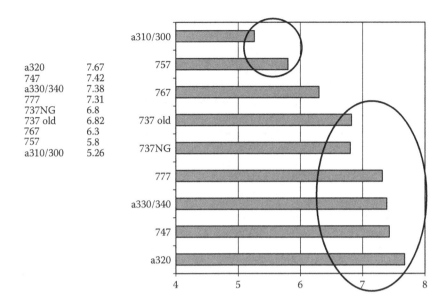

a320	7.67
747	7.42
a330/340	7.38
777	7.31
737NG	6.8
737 old	6.82
767	6.3
757	5.8
a310/300	5.26

FIGURE 3.13   The comfort scores for different aircraft types. The scores within the oval and the circle did not differ significantly ($p < .05$, t-test).

Boeing 777 group. Also, within one group we see differences. The Boeing 767-400 scores significantly higher than the 767-200. This could be caused by the fact that the 767-400 looks newer.

On the other hand, the various versions of the Boeing 747 were not significantly different from each other, while here the 200 series is much older than the 400 series. The difference between airlines in this study has more influence than the aircraft type.

Similarly, in the smaller planes, few significant differences were shown, except that the Embraer 170-195 series scored significantly higher (Figure 3.14) and the CL600-700 series significantly lower. We would expect the DHC8, ATR42/72, and Fokker50s to score lower as these are turboprop planes making more noise and flying on a lower altitude, which causes more turbulence, but these factors were apparently not important enough to influence the comfort score. They do not differ significantly from the rest of the planes.

Perhaps the clearest conclusions that can be drawn from this study is that factors like leg room, hygiene, and staff have more influence on comfort than the aircraft model. Also, the configuration and age of the plane are of more importance than the type itself for the comfort rating.

DIRECT VERSUS NO DIRECT FLIGHT

The statement that an indirect flight would be less convenient could make sense. A transfer means extra walking, waiting for the gate, and sometimes additional queues due to customs could be experienced as inconvenient. We did split the direct

FIGURE 3.14 The interior of Embraer 190 is one of the best rated planes. We should be careful with this conclusion because it is a rather new airplane and the fact that there is no middle seat also influences the average comfort score.

and indirect flights, and there is no relationship with comfort (r = .01). Probably, expectations play a role here. The moment you have booked the flight, you are often aware of this situation.

REFLECTION

Comfort is an important issue in flying. It has some influence on the fact of whether a passenger will book with the same airline again and, thus, is worthwhile to pay attention to. We are at the beginning of understanding how the comfort experience is built. Soft factors like crew attention and preflight experience play a role, but physical characteristics, such as the seat characteristics, do as well. There also are processes influencing each other. More noise is not mentioned and is a separate problem. The previous chapter shows that it could influence aspects like awareness of swollen feet. This study shows that passengers are aware of the leg room, hygiene, the crew attention, and the seat, which all have clear relationships with the comfort experience and have priority in improving the aircraft interior.

The study also has some drawbacks as only trip reports are used. It could be that this is a selection of the real travelling population. However, the study of Blok, Vink, and Kamp (2007) showed that trip reports and questioning after a flight did not lead to different outcomes. Another drawback is that the effect of light, noise, and other influences of the environment that passengers do not notice explicitly are not recorded. It could be that these factors influence comfort, but for these specific factors other forms of research are needed. Experimental studies where only one factor (e.g., light or noise) is changed and the effects on passengers are measured and compared with a control group. On the other hand, if passengers select a flight they only take the issues into account that come into their minds and, therefore, paying attention to leg room, hygiene, crew, seat, and expectations is a way to increase company revenues.

REFERENCES

Blok, M., P. Vink, and I. Kamp. 2007. Comfortable flying: Comfort in aircraft interiors seen through the eyes of the passengers (in Dutch). *Tijdschrift voor Ergonomie* 32 (4): 4–11.

Brauer, K. 2004. Convenience, comfort, and costs. Presentation at the aircraft interior EXPO.

Brauer, K. 2010. Redesigning the passenger experience. Paper presented at Stanford Graduate School of Business, Stanford, CA, April 20.

Oborne, D. J. 1978. Passenger comfort—An overview. *Applied Ergonomics* 9: 131–136.

Pine, B. J., and J. H. Gilmore. 1999. Experience economy. Boston: *Harvard Business School Press.*

Richards, L. G., and J. D. Jacobson. 1977. Ride quality assessment III: Questionnaire results of a second flight programme. *Ergonomics* 20: 499–519.

Schifferstein, N. J., and P. Hekkert. 2008. *Product experience.* Amsterdam: Elsevier,

Vink, P., C. Bazley, I. Kamp, and M. Blok. 2011. Possibilities to improve the aircraft interior comfort experience. Forthcoming. *Applied Ergonomics.*

Zenk, R. 2008. Objektivierung des sitzkomforts und seine automatische anpassung. PhD thesis, Technical University, München.

4 New Demands for Aircraft Seats Based on Recent Research

Overviews: Research shows that a seat provides a higher comfort level if it:

* Accommodates reading in the seat with the backrest tilted rearward.
* Accommodates varying the seating posture.
* Accommodates various body sizes.
* Provides an ideal pressure distribution (perhaps by an intelligent seat that senses pressures and adapts itself).
* Has no shear forces on the seat.
* Enables doing different activities comfortably in the seat.
* Provides a "wow" experience at first sight.
* Includes an option to lift the feet off the ground.
* Provides the feeling that the backrest follows the curve of the body.
* Is adjusted easily (perhaps through the use of electronics).

A final test of any design with real end users is absolutely necessary because the interaction between factors is difficult to predict and there are factors that remain little understood or unknown.

USING RESEARCH FOR SEAT DESIGN

Purchasing, marketing, engineering, or designing an aircraft seat is far from simple. First of all, the aircraft seat is uniquely challenging. It is challenging because there is no other type of seat used by such a wide variety of body types in such a restricted space for such long periods. Additionally, the regulations governing aviation safety are very strict and absolutely compulsory.

On the other hand, using the latest knowledge to create unique customer value is a good way for an airline or supplier to distinguish itself from its competitors. This chapter will review some research findings regarding sitting and seats that may be helpful in creating a new product and distinguishing one airline from other airlines or manufacturers.

SEAT DESIGN AND HEALTH

Some years ago, a number of manufacturers of office seating stated that their chairs prevented back complaints. Supporting this point of view is not difficult as Nordin (2004) showed in her review of high-quality epidemiological studies that suggested that seating itself is not a risk factor for back problems. In those epidemiological studies, large groups of individuals were followed and comparisons were made between groups that sit frequently and others that did not sit frequently. These studies measured whether back problems developed over a span of years. The many studies reviewed showed that sitting by itself is not a risk factor for back pain. As a result, a seat cannot be considered to prevent back complaints. For neck/shoulder pain, the situation is different. Some studies show a relationship between frequent sitting and having neck/shoulder complaints (Ariens, 2001).

Years ago, seat manufacturers stated that their seats enabled healthy upright sitting with the back and upper legs positioned at a 90 degree angle from one another. This point of view is hard to defend as there is evidence that a backward leaning posture reduces the load on the lumbar back (Wilke et al., 1999; Figure 4.1). Zenk (2008)

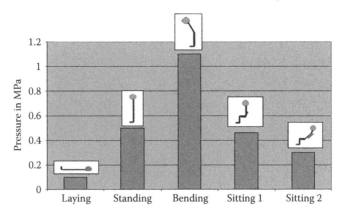

FIGURE 4.1 Pressure on the intervertebral disc between L4 and L5 in MPa in different postures, according to Wilke et al. (1999). A higher pressure is correlated to a higher loading of the spine.

also showed that a posture with the trunk supported and positioned leaning some-what backward resulted in the least discomfort while driving a car. Groenesteijn et al. (2009) showed that, for reading while seated, an office chair with a reclined backrest resulted in higher comfort ratings than one with an upright backrest. The view that it is healthier to sit upright with the knee and hip bend in a 90 degree angle is not supported by scientific research.

To make it even more complex for seat developers, the number of experts in the field of musculoskeletal loading promoting a seat that stimulates movement is growing. Nordin (2004) states, based on her review of epidemiological research, that sitting itself is not a risk factor. However, prolonged sitting in a restricted posture is a risk factor. The restriction of posture does increase the chance of back problems, which is an argument showing the need for movement while seated. There are other scientific studies (e.g., Dieën, Looze, and Hermans, 2001) that indicate that more dynamic sitting and more variation in posture (Lueder, 2004) is better for the back. For instance, Dieën, de Looze, and Hermans found that the length of the human body increased significantly more after sitting in a chair that facilitated movement of the body than sitting in a fixed chair. Normally, the length of the spine declines during the day due the load of the upper body and it recovers during the night. In this case, the movement while seated had a recovery effect similar to, although less than, that resulting from an evening's rest. The change in posture was possible because the chair had a movable seat and backrest.

With this in mind, it is interesting to study how passenger movement can be facili-tated in an airplane seat. Of course, this is not easy. When driving a car, we meet a similar challenge. It is not wise to move the body much when driving because atten-tion needs to be paid to the traffic. The automobile manufacturer BMW addressed this issue by adding a lightweight massage system (Franz, 2010). The movement pattern of this 60 gram pneumatic system (Figure 4.2) was developed in such a way that the driver did not get sleepy. An experiment with 20 drivers driving for approxi-mately 90 minutes showed that activity of the shoulder muscles was significantly decreased when using the massage system compared with driving without the sys-tem. This was recorded by placing electrodes on the muscles that record muscle tension (electromyography or EMG). The comfort was highly appreciated and there was no distraction. Also, the pressure in the intervertebral discs (the fluid-filled discs in the spine that link the different vertebrae) varied when the massage system was on, indicating that this specific massage pattern promotes fluid transport in the discs, which could have a recovery effect.

AIRCRAFT SEATS SHOULD FIT

In addition to a backrest tilting backward and a feature facilitating movement, a seat should be comfortable for small Asian women and tall Dutch men. There are guidelines that are helpful in this area. The Web site www.dined.nl contains anthro-pometric data (data on human body sizes) for different regions in the world. As an example, seated hip widths can be found on this site. The P95 hip width for males between 31 and 60 years old living in the Netherlands is 440 mm (Table 4.1). P95 means that 95 percent of the males have a smaller hip width than the 440 mm. So,

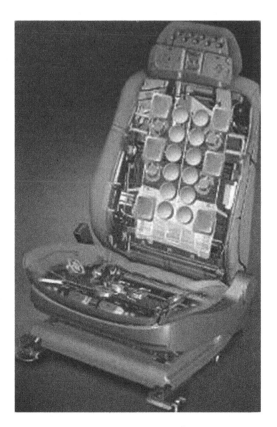

FIGURE 4.2   The BMW seat with the massage system of Franz (2010). The 12 round and 6 square elements are inflatable and generate the massage effect.

TABLE 4.1
The P95 Hip Width of 31- to 60-Year-Old Males in Different Regions of the World While Sitting

	mm	in.
North India	330	13.0
Japan	348	13.7
Australia	370	14.6
Middle East	370	14.6
Latin America	388	15.3
North America	394[a]	15.5
Central Europe	404	15.9
Netherlands	440	17.3

[a] According to CAESAR data (2000), the P95 U.S. male hip width is 436 mm.
Source: www.dined.nl (accessed July 1, 2010).

TABLE 4.2

The Thickness or Space Required to Get a Comfort Grade for a Seat According to Boeing Guidelines

		Grade			
		A	B	C	D
Leg room space					
1	Thickness at knee height (24.9 in. above floor)	<1"	1"–2"	2"–3"	>3"
2	60° shin clearance (from SCRP)	<0.8"	0.8"–1.7"	1.7"–2.5"	>2.5"
3	45° shin clearance (from SCRP)	<0.5"	0.5"–1.2"	1.2"–1.9"	>1.9"
Back and shoulder space					
4	Lumbar depth	<0.5"	0.5"–0.8"	0.8"–1.1"	>1.1"
5	Shoulder obstruction height	>25.8"	24.8"–25.8"	23.7"–24.8"	<23.7"
Working, eating, and visual space					
6	Upper back thickness	<1.5"	1.5"–2.5"	2.5"–3.5"	>3.5"
7	Headrest thickness	<1.5"	1.5"–2.8"	2.8"–4"	>4"
8	Space between seat backs (27.6" above SCRP)	<4"	3"–4"	2"–3"	<2"

Note: See also Figure 4.3. The numbers in the left column correspond to that figure.

in the Netherlands 95 percent of the male passengers between 31 and 60 years old fit in a seat that is 440 mm (17.3 in.) wide. The hip width in Japan and northern India is smaller and, in those countries, a seat with a width of 348 mm and 330 mm (13.7 in. to 13 in.), respectively, is sufficient for 95 percent of the population.

Translating these data to a seat means that to accommodate 95 percent of Dutch men the distance between two armrests that reach all the way down to seat cushion should be 440 mm (17.3 in.). However, variation in posture is difficult if the hip is stuck between the armrests. As a result, more space is needed to sit comfortably for a longer time. This, of course, only relates to seat width. Relevant data for other seat dimensions can be found at www.dined.nl.

In the airline industry, the Boeing spatial comfort guidelines are often used; these guidelines are based to a large degree on selected data from CAESAR (2000). Various attributes of a seat, particularly thicknesses that take available space away from the passenger, can be rated as an A, B, C, or D. The guidelines are illustrated in Table 4.2 and Figure 4.3. The numbers of the first column in Table 4.2 correspond to the numbers in Figure 4.3. An attribute rated as an A, being relatively thin, gives the most space to passengers. As an example, a thickness at knee height of less than one inch is rated as an A and results in relatively more space available for the passenger sitting behind this seat. However, it also is dependent on the pitch (the distance between like points on seats in successive rows, e.g., between the back of one seat and the back of the seat immediately behind it).

However, many things that have a significant impact on comfort are not addressed in the Boeing guidelines, in large part because they are less easily quantified. The ability to vary posture, the appearance, the cushion characteristics, the three-dimensional

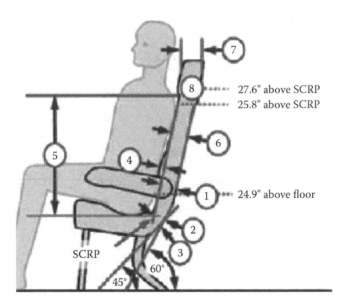

27.6" above SCRP
25.8" above SCRP

24.9" above floor

SCRP

60°

45°

FIGURE 4.3   The Boeing guidelines for seat comfort.

form, and the support provided to the different parts of the body are important as well. Besides, not every part of the body is equally important. In the Chapter 3, we saw that the hip-to-knee space is seen as very important to passengers. From the literature, it is known that the pressure distribution in the seat pan is related to comfort (Looze, Kuijt-Evers, and Dieën, 2003). And, we also want the seat to feel comfortable during sleeping, eating, watching in-flight entertainment (IFE), working, and reading. These activities demand different postures that should be accommodated comfortably by the seat.

PITCH WATCHERS

Sometimes passengers mentioned that "this seat has a good pitch" in their trip report. They even stated that at Web sites the comfort can be found by looking at the pitch. These "pitch watchers" can find seat pitches on sites, such as seatguru.com, easyair-plane.com, airlinequality.com, and many more. They are presuming that a larger pitch necessarily provides more legroom and as a result more comfort. This is only true if the seats in the airplanes being compared are of the same thickness. The thickness of the back support influences the hip to knee space. A pitch of 33 in. and a 3-in.-thick backrest gives less knee space than a pitch of 31 in. and a 1-in.-thick backrest. A major goal of the Boeing guidelines is to grade the thickness of regions of seats critical to comfort so that their impact on hip to knee distance and other fore and aft distances can be readily understood. The letter grade increments for thickness measures are even specified so that a one letter grade improvement results in giving the passenger one additional inch of fore and aft space in the selected region.

To appear knowledgeable, passengers and others working in the airline industry do well to say, "The leg room feels like there is more pitch ...."

DESIGNING AN AIRCRAFT SEAT IS DIFFICULT

In addition to the above mentioned demands, a seat also should be lightweight to meet the environmental requirements and thin to be able to have as many passengers as possible aboard at the same comfort level. It might seem impossible to meet all the requirements, but it is not. The airline industry has been able to increase comfort ratings in the past. Newer planes are rated significantly better with regard to comfort than older ones (see Chapter 3). However, we also know from the comfort theories that passengers compare their sense of comfort with their experience in other airplanes and in other fields. So, when the automotive industry is constantly improving, other sectors have to follow suit. The comfort theory also shows that experiencing more than expected increases the comfort rating. To experience more than expected, several attributes or goals could be of help and might include:

- An ideal pressure distribution
- An ideal backrest angle
- Prevention of shear forces
- Massage systems
- Seats that fit to specific activities
- Seats that adapt themselves (smart seats)
- Possibilities to have the feet off the ground
- Wow experiences
- Seats that suggest comfort visually

In the next two chapters, these elements will be described in more detail.

IDEAL PRESSURE DISTRIBUTION

Of all measurement methods indicating (dis)comfort, pressure distribution is one of the measurements that has the most clear relationship with discomfort, according to a review of the scientific literature of Looze, Kuijt-Evers, and Dieën (2003). A better pressure distribution can reduce discomfort. Even a relationship between pressure distribution and neck and back complaints has been established. Addressing discomfort according to a special procedure can result in a design that reduces the chance of sick leave due to neck and back complaints. Hamberg et al. (2008) showed that a lower level of discomfort does significantly reduce the chance of neck and back complaints. She followed approximately 1,700 subjects for three years and the participants with the higher discomfort developed more complaints three years later.

There is no hard evidence revealing precisely what pressure distribution is healthy and comfortable. Some authors see 60 mm Hg as the absolute maximum for pressure

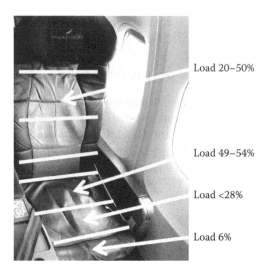

Load 20–50%

Load 49–54%

Load <28%

Load 6%

FIGURE 4.4   The ideal pressure distribution for driving a premium car, according to Zenk (2008) and Mergl (2006). If we assume the posture in aircraft seats is similar, this is the ideal pressure distribution.

(Conine and Hershler, 1994), as much of this fluid transported through the vessels is blocked above this pressure. However, around the ischial tuberosities (the two bones in the buttock you feel when you sit on a wooden chair) this pressure can be higher. The maximum acceptable pressure also is strongly dependent on the duration of the sitting activity and the variation of posture while seated. While our understanding of the relationship between pressure and comfort may be incomplete, there is no question that pressure is related to discomfort. Zenk (2008) showed in his PhD thesis that a reclined backrest results in a high comfort score. He also showed that some support under the front of the legs, spreading the loads and thereby reducing pressure, reduces discomfort in the long run. After 2.5 hours of driving, a seat designed with this detail in mind was still rated as comfortable. In Figure 4.4, a pressure distribution providing a low level of discomfort for a premium car seat is shown.

For aircraft seats, this profile still has to be established, but it may be somewhat similar since some postures employed by airline passengers resemble the posture of an individual driving a car. It is a bit more complex for an aircraft seat because the activities and, therefore, the postures vary more. In driving a car, the activity is clear. By contrast, in airplanes, passengers have dinner, sleep, and watch movies, activities that all have rather different comfortable postures. An aircraft seat should result in a desirable pressure distribution when the occupant is in any of these postures, which makes a comfortable seat design far more complex. Apart from the comfort experience, which is often increased by a better pressure distribution, there are also indications in the literature that a good front leg support reduces the pressure in the intervertebral discs (Zenk, Franz, and Bubb, 2010). The pressure under the body is also better distributed in these cases.

Interesting for business class seats is the automatic seat adaptation system described by Zenk (2008). In these seats, sensors record the pressure and the seat automatically adjusts itself, finding the ideal pressure distribution (see Figure 4.4).

SEATING AND SHEAR FORCES

Tilting the back support rearward while keeping the seat bottom horizontal increases the tendency to slide out of the seat in a forward direction. However, this forward sliding force is counteracted by friction between the seat and the occupant preventing him/her from sliding out of the seat. This forward force on the seat is called a shear force, which is a topic of debate among scientists regarding its relevance to health issues. However, it is clear that this force exists and has a relationship with comfort. The question is how to cope with it in designing a seat. It can be ignored of course: "If the passenger wants to slide out of the seat, it's not our business." Or, a seat can be designed in such a way that the shear forces are low. Extreme shear forces can lead to discomfort and if they are maintained long enough or, if the forces are strong enough, they could lead to a decubitus ulcer (Goossens and Snijders, 1995). It could also lead to lipoatrophia semicircularis, a circumferential furrow in the skin of the thigh (Goossens, 2001). However, there is a debate among scientists regarding this relationship. Perhaps discomfort prevention is the main reason to prevent shear forces. Shear forces can be reduced by tilting the seat pan with the front of the seat upwards. Figure 4.5 shows what the ideal seat pan angle should be for any given backrest angle to prevent shear forces on the seat based on a biomechanical model

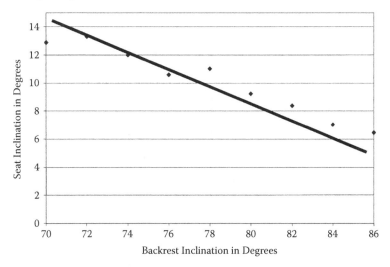

FIGURE 4.5 Shown is the seat and backrest inclination during sitting without shear force on the seat. The continuous line represents the prediction of a biomechanical model. The dots represent the mean of the measured inclinations on healthy subjects (Goossens and Snijders, 1995).

and a measurement at test subjects. It is a useful figure to consult to determine if there is an unacceptable shear force and to guide the redesign if there is an unacceptable shear force.

COMFORT AND SEATING

It is important to realize that comfort is not only influenced by pressure or physical characteristics. Kuijt-Evers (in Bronkhorst, 2001) showed that 49 experienced office workers evaluated one out of four office chairs negatively based on the visual information. The four seats were exactly the same physically, only the colours differed. Three seats were light coloured and one was brown. The first impression was that the brown coloured seat would be less comfortable. The first seating experience after this visual inspection also resulted in lower comfort ratings. The brown chair, however, was evaluated as positively as the others after using it for more than an hour. The fact that expectation, first sight, and individual differences play a role in the comfort rating means that expectations and first impressions must be carefully controlled in experiments. Of course, the seat should have good pressure distributions, appropriate dimensions (anthropometry), and be easily adjustable, but the emotional part should be taken care of as well. Measuring the first impression can be helpful in this case. There are useful methods available for recording the first impression, such as measuring the activity of the muscles in the face (e.g., musculus zygomaticus, see the paragraph below on comfort and "wow") and the face reader to see the first reaction it has on human beings. The face reader is software analysing a digital face picture or movie, making it possible to record emotions.

Another aspect in testing is that the comfort experience is activity or task dependent. Groenesteijn et al. (2009) showed that the comfort experience in a seat while reading was influenced by the angle of the backrest. Individuals often have the backrest positioned rearward while reading and an increased range of motion of 8 degrees resulted in better comfort ratings. As a result, comfort measurements are needed of occupants performing each specific task for which the seat will be used. Activity specific designs are needed (Figure 4.6).

The fact that it pays off to design seats that fit the relevant activities is shown in a project in which train seats were designed for the Long Island Railroad (Bronkhorst and Krause, 2002; Vink, 2005). After a search on the Web (among allied institutes) and in the literature, it appeared that there was no knowledge available on how people sit in a commuter train and what their main activities were. Therefore, a project was undertaken to observe the behaviour of passengers on the routes the new train will serve. Measurements were taken to monitor the passengers' activities, posture, size, and movements. From 1,700 observations, the four most frequent tasks were selected as well as the essential anthropometric characteristics. In this way, a seat could be designed that best fit these four activities and during the development process several tests were done with real subjects to check whether these activities could be performed while sitting comfortably. Activities mentioned by passengers as being important also were tested (Figure 4.7). It appeared that, in the end, 83 percent of the passengers preferred the resulting seat to current seats.

FIGURE 4.6 The test of a train seat during the activity reading.

FIGURE 4.7 One of the experiments done with a mock-up to test the in- and egress for the railway seats.

Such an experiment is not easy. Simply asking: "Is it comfortable?" will not give information needed to guide a redesign and does not provide a comparison to other seats.

Because the comfort experience of human beings is still somewhat unpredictable and is dependent on many factors, studies with real subjects are always needed. More information on foam characteristics, forms, and safety are becoming available in virtual (comfort) models (Franz, 2010), but these systems are not far enough along in development to replace studies with real subjects. These studies with real subjects can be done in an early phase (looking at the design) or with prototypes or in virtual reality. Many successes in projects where full-scale objects, mock-ups, and (virtual) prototypes have been used are reported (Davies, 2001; Franz, 2010). Of course, these studies need to be done in an as natural a setting as possible to exclude environmental influences.

SPECIFIC DYNAMIC SEAT CHARACTERISTICS

In the office seat market, seats are available with specific dynamic seat characteristics, such as seat pans that float a little or rotate along a longitudinal axis. To study the effects on the human body of these features, extensive research was done by the BGIA (Berufsgenossenschaftliches Institut für Arbeitsschutz, an institute for occupational safety and health) in Germany (Ellegast et al, 2008). In this study, four more expensive chairs with special dynamic features were compared with a now often used office chair that has a synchro mechanism, which means that the seat pan and backrest can move in the sagittal plane (front–backward), but the movement of both parts is coupled to one another.

The specific dynamic feature of the first chair is a small electromotor that activates the movement of the seat pan alternately and constantly 0.8 degrees to the left and to the right, five times per minute. The main dynamic element of the second chair is the suspension system of the seat pan that allows movements in the horizontal plane. Comparable to a swing, the seat pan of the third chair is fixed on a pendulum so that it can be moved freely in all directions, and the last tested chair includes a three-dimensionally moveable joint that allows the seat pan to move freely in all directions. Muscle activity, pressure distribution, human movement, and posture and comfort were recorded in a laboratory and in the field with 30 participants. These participants had to perform different tasks, like telephoning, computer work, and filing. Almost no differences between the seats were found on human body movement, posture, and muscle activity. On the other hand, the effects of the different activities (telephoning, filing, or computer work) were large. This shows that it is probably wiser to invest in seats suited to the various office tasks than in a seat that has complex features intended to have the human body move more. Some difference in the comfort experience was found between the seats, which was also dependent on the tasks (Groenesteijn et al, 2010). There are some indications that telephoning was most advantageous in the first chair, computer work was best in the "swing system," and filing in the last chair, which moved like sitting on a ball. If we translate this research to aircraft seating, it is probably wise to focus on varying the posture by changes in activity and introducing a synchro mechanism is probably enough. It also indicates that it is preferable to have task-specific areas (see Chapter 3).

COMFORT AND "WOW"

One should ideally have a product that generates a "wow" experience. This may sound a bit obvious, but the techniques to measure this "wow" experience at first sight are developing fast, and it is worthwhile to use these in product design. There are studies that show that the more "wow" experience we have the higher the comfort rating (Franz, 2010). This is not strange, as we already saw that expectations and first sight play a role in the comfort experience. A seat should not only feel comfortable over the long term, but also look comfortable. In fact, a design should

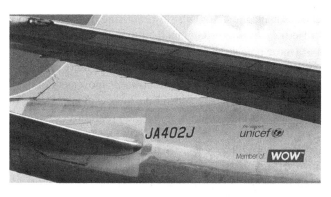

FIGURE 4.8   The "wow" factor is also used in marketing.

elicit a high comfort feeling at first sight, higher than expected. A "wow" experience should be generated. This "wow" reaction can be measured. As an example, Microsoft used the activity of the musculus zygomaticus (Hazlett and Benedek, 2005). These are muscles (major and minor) that play a role in laughing. In the event of a "wow" experience, these muscles are more active even before we are aware of the "wow" experience.

"Wow" is often used in marketing (Figure 4.8). Many methods measuring "wow" experiences use questionnaires and interviews and are of a subjective nature. A fundamental problem in these studies is that, in using the subjective method, a verbalization phase has passed that is a conscious activity. In this way, the preconscious "wow" activity is not measured. The measurement technique of Desmet, Porcelijn, and van Dijk (2005) avoids the verbalization phase by showing pictures or movies to test subjects. However, choosing between movies or pictures is also a conscious activity. The advantage is that we are not now restricted to words. Riseberg et al. (1998) found a relationship between blood volume pressure, skin conductance resistance, and muscle tension on the one hand and situations that provoked negative emotions. However, blood pressure and skin conductance only indicate the arousal level and, therefore, are less precise. The study of the brain with methods like functional

MRI (magnetic resonance imaging) also is promising, but the current theory is too rudimentary (Tierney, 2004). EMG (a method to study muscle activity) looks more promising. It is a more precise and sensitive method to measure changes in the facial expression, which also precedes the conscious phase. Activity of the musculus zygomaticus major (a muscle that plays a role in upwards movement of the corner of the mouth) and the musculus corrugator superclii (a muscle of importance in contraction of the upper part of the eyelid) could be connected to positive emotions as these are connected to smiling.

Hazlett and Benedek (2005) tested whether EMG of these muscles was related to the subjective self-report. Several desktop screens, such as backgrounds, menus, and icon organizations, were shown on a VDU (visual display unit) screen and the reaction was measured by EMG and a self-report. It appeared that the participants' self reports were strongly correlated to the physiological measurements.

Also, regarding this aspect, it is possible to test various versions of aircraft interiors and aircraft seats and select the versions that evoke the best "wow" emotion. Various pictures can be shown (such as those we see on the Internet because more sales are made on the Internet) and the immediate reaction of purchasers and end users can be studied. It is useful to predict Internet sales and to know how comfort at first sight will be evaluated. In the end, of course, sitting in the seat should give the end user more comfort than expected to maintain the positive emotional experience. Therefore, it is also important to know how current seats are experienced since new seats should at least have this comfort level.

FEET OFF THE GROUND

An interesting finding in developing a lounge chair was the fact that, to achieve a comfortable position watching a screen, subjects have their feet off the ground (Rosmalen et al., 2009). Rosmalen and colleagues observed and videotaped subjects in their natural environment. Researchers were not in the room, but the video camera recorded the behaviour of individuals at home watching a screen (Figure 4.9). It appeared that most of the time the feet were off the ground. Rosmalen's group also asked other subjects to write down the most comfortable position followed by a group session. She also performed a laboratory experiment in which participants tested various positions.

All experiments clearly showed a variation in postures, as mentioned earlier. All studies showed that having the feet off the ground is preferred for a comfortable posture. In many business class seats, there are possibilities to have the feet off the ground, which is a good choice if we want to accommodate natural behaviour. Even some economy class aircraft seats have the possibility to have the feet off the ground, but further innovations in this area could be attractive as the current seats do not always allow a variety of postures. Also, the natural positions human beings take while seated are often not supported. That is an understatement if we look at the results of the survey. Watching the natural behaviour of people while sitting is a good inspirational source for innovations in the aircraft interior. Research techniques from the field of anthropology can be very useful in this area.

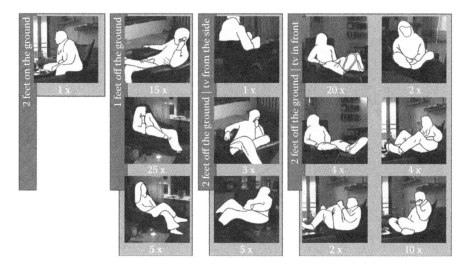

FIGURE 4.9 Observed comfortable sitting postures while watching a screen. In 45 of the observed cases, one foot was off the ground and, in 51, both feet were off the ground (Rosmalen et al., 2009).

BACKREST ANGLE

Much is written about the most comfortable backrest angle. Most of these studies are of a discussion or visionary nature. Some studies really did measure the posture and some are a guideline based on data of others. Rosmalen et al. (2009) described 110 degrees as the optimal backrest angle for watching a screen (110 degrees with respect to the horizontal, in fact, 20 degrees backward, with respect to a vertical), while Nathan-Roberts et al. (2008) describes 100 degrees. One difference was that, in the work of Rosmalen and colleagues, the seat pan was able to be tilted in various positions. Park et al. (2000) observed sitting postures while driving and found a mean backrest angle of 117 degrees varying from 103 degrees (min) to 131 degrees (max). Sitting in a car seat, Anderson et al. (1974) and Hosea et al. (1986) both found 120 degrees as the position with the lowest muscle activity. Harrison et al. (2000) describe that, in case of a 5 degree tilted seat, a 100 degree trunk position is preferable, as a more backward backrest would result in too much neck flexion while watching the road. The position of the back depends strongly on the task (Dunk and Callaghan, 2005) and, therefore, it is difficult to come up with one number for the ideal backrest position. However, the study of Wilke et al. (1999) and Rosmalen et al. (2009) indicate that a slightly backward tilted backrest in the range of 100 to 120 degrees is preferable.

It is not only the backrest angle that determines the comfort. Kroemer et al. (2001) also describes the importance of the form of the seat. The preferred two-dimensional form according to Kroemer et al. (2001) is shown in Figure 4.10. Also, in the other dimensions, the form is important, it is even sometime dependent on the culture. Vercaygne-Bat (2008) showed that German drivers like more the wings in the seat, while French drivers preferred more of an equally formed seat.

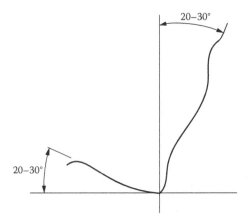

FIGURE 4.10   The shape of the seat in the Kroemer et al. (2001) study.

FIGURE 4.11   Two seat controls for business class seats.

SEATING AND ELECTRONICS

To find the optimal posture, seat electronics can be useful to steer parts of the seat. In many business class seats, electronics have been introduced (Figure 4.11). However, one difficulty is that the seats are often not adjusted to an optimal position because the controls are difficult to understand. Office chairs also are often not properly adjusted (Vink et al., 2007). To find out how many office workers adjust their chairs, 336 office workers in Spain and the Netherlands were observed and questioned on whether they adjust their chairs. It appeared that 24 percent of 236 Spanish office workers and 61 percent of 100 Dutch subjects never adjust their chair. If the chair is adjusted, it is most commonly the seat height that is adjusted. Except for the seat height, other adjustment possibilities are not used by the majority of the study population. Reasons for not adapting could be awareness, complexity of the control system, and expected effects. This problem was also discovered in seat adjustments in cars. Therefore, Zenk et al. (2011) developed an automatic seat adjustment system for the automobile manufacturer, BMW. First, the ideal pressure distribution was established using 40 subjects driving for 2.5 hours in various pressure distributions (see Figure 4.1). Pressure sensors were positioned in the seat to measure the pressure at various locations and an algorithm was developed that could electronically adjust the various parts of the seat. In use, the system searches for the position that comes closest to the ideal pressure distribution. Again tests were done with this seat adjusted in such a way that it resulted in the ideal pressure distribution. The discomfort of drivers in the seat that automatically provided the ideal pressure distribution was compared with the discomfort experienced by drivers who could adjust the seat themselves. Forty subjects drove 2.5 hours in both conditions. The results of this study are very telling. There was significantly less discomfort in the automatic seat condition, with 95 percent of the test subjects rating the system as very good.

This procedure can be applied to aircraft seats as well. It is possible to warn passengers of the ill effects when they are seated too long in one position. There are now office seats available with pressure sensors (Kuijt-Evers et al., 2007). She also showed that it is possible to characterize the position of the office worker with these seats and measure how long subjects are positioned in one posture giving more possibilities for advice to passengers. Dauphin, a manufacturer of office seats, uses a similar system to warn the users if they do not touch the backrest.

Also, for the controls that are used to adjust a seat, tests with real subjects are needed to arrive at systems that are easy to use so that passengers will feel comfortable. From the survey in this book, it was clear that some people did not adapt their chair because they felt stupid if they had to ask the flight attendant how to do it. Trying other positions and making mistakes and being seen by other passengers in their attempt to adjust the seat and doing it wrong were also mentioned. This is another reason to make systems user friendly.

Despite the fact that we know what needs to be done to support passengers in making the adaptation logical and despite the fact that we have systems that record the position and pressure and adapt the seat automatically based on that information, there are still too many uncertainties to making a system that gives passengers the optimal comfort experience.

There are still too many factors that influence the behaviour of passengers to be able to predict how these systems will be used, making tests with real subjects a necessity.

OTHER FEATURES: HEADRESTS AND MASSAGE

Several studies show that other features could increase comfort as well. As mentioned above, Franz (2010) showed that the muscle activity is reduced due to massage and the comfort is increased. He also showed that the comfort can be increased by a headrest and a neck rest. For providing comfort in a car, it is important to have a softer foam at the neck and a more stiff foam for the contact area with the head. Perhaps this is true for aircraft as well. If we go back to the trip reports, the firmness of the foam is not often mentioned by passengers. Passengers report that often the headrest has no side support and the headrest is too far back. The latter is also caused by the fact that the anthropometric data on the position of the back of the head with respect to the back of the shoulders are lacking in different positions. Franz (2010) measured 35 subjects from P5 to P95 to establish the ideal headrest position and used this in designing a new headrest.

For sleeping, a flat bed seems the ideal solution. Often in the business class, flat, inclined beds are offered to attract passengers. However, the passengers in the trip reports complain about sliding out of the seat. If the bed is not horizontal, passengers slide out of their seats and a curved seat could even be better for the sleeping comfort.

The lumbar support also could contribute to comfort improvement (Lueder, 2004). Already in 1974, Anderson et al. (1979) found that in the upright sitting posture a lumbar support reduces the muscle activity. However, the ideal form is strongly dependent on the posture, activity, and person making the lumbar support design a challenge.

For all of these situations, it is wise to study the literature, use past experience, available virtual models, and eventually test various versions with real test subjects in as naturalistic setting as possible.

OPPORTUNITY FOR DESIGNERS

There are no aircraft seats in development now that fulfill all the requirements mentioned in this chapter. There is no seat with sensors that gives a "wow" sensation at first sight, has a higher comfort than expected, fits to the activities (or is adjustable to the task), has almost no shear forces, stimulates movement, has the ideal pressure distribution, enables seating with ideally formed backrest, and is made so that one sits with the feet off the ground.

This is strange, as there are many aircraft seat manufacturers, and a lot of research and development is done in this area. However, most research concerns testing the seat compared with other seats (e.g., previous versions) after development to show how good the new seat is. Probably no one, as of yet, has studied the literature and used it in its design. Also, meeting the safety demands is already so difficult that there is not much room left for extra comfort development. The challenge becomes even larger as seats also should be lightweight. And, of course, some knowledge has become available only recently and now it is the right time for new innovations.

REFERENCES

Anderson, G. B. J., R. W. Murphy, R. Örtengren, and A. L. Nachemson. 1979. The influence of backrest inclination and lumbar support on lumbar lordosis. *Spine* 4: 52–58.

Anderson, G. B. J., R. Örtengren, A. Nachemson, and G. Elfstrom. 1974. Lumbar disc pressure and myoelectric back muscle activity during sitting. IV studies on a car driver's seat. *Scandinavian Journal of Rehabilitation Medicine* 6: 128–33.

Ariens, G. A. M. 2001. Work-related risk factors for neck pain. PhD thesis, Vrije Universiteit, Amsterdam.

Bronkhorst, R. E., and F. Krause. 2002. End-users help design mass transport seats. In *Human factors in seating and automotive telematics (SP-1670)*. SAE World Congress, Detroit, MI, March 4–7. Warrendale (PA): SAE: 1-6. SAE Technical Paper Series 2002-01-0780.

Bronkhorst, R. E., L. F. M. Kuijt-Evers, R. Cremer, J. W. van Rhijn, F. Krause, M. P. de Looze, and J. Rebel. 2001. Emotion and comfort in cabins: Report TNO, Hoofddorp, Publ.nr. R2014871/ 4020054.

CAESAR. 2000. Data from the Civilian American and European Surface Anthropometry Resource Project—CAESAR.

Conine, T. C., and C. Hershler. 1994. Pressure ulcer prophylaxis in elderly patients using polyurethane foam or jay wheelchair cushions. *International Journal of Rehabilitation Research* 17: 123–37.

Davies, R. C. 2000. *Using virtual reality for participatory design and brain injury rehabilitation.* Lund, Sweden: Lund University (Ph.D. thesis).

Desmet, P. M. A., R. Porcelijn, and M. B. van Dijk. 2005. How to design wow. Introducing a layered-emotional approach. In *Proceedings of the Conference Designing Pleasurable Products and Interfaces*, ed. S. Wensveen. Technical University/Eindhoven, the Netherlands, pp. 71–89, 2005.

Dieën, J. H. van, M. P. de Looze, and V. Hermans. 2001. Effects of dynamic office chairs on the low back. *Ergonomics* 44: 739–50.

Dunk, N. M., and J. P. Callaghan, 2005. Gender-based differences in postural responses to seated exposures. *Clinical Biomechanics* 20: 1101–1110.

Ellegast, R., K. Keller, R. Hamburger, H. Berger, F. Krause, L. Groenesteijn, M. Blok, and P. Vink. 2008. Ergonomische untersuchung besonderer büroarbeitsstühle. Sankt Augustin, BGIA.

Franz, M. 2010. Comfort, experience, physiology, and car seat innovation, PhD thesis, Delft University of Technology.

Goossens, R. H. M. 2001. Lipoatrophia semicircularis, een hypothese. In *Lipoatrophia semicircularis, Informatiebrochure samengesteld naar aanleiding van het wetenschappelijk symposium ribbeldijen en de relatie tot beeldschermwerk* (14-15). Brussels: Ergoclinic.

Goossens, R. H. M., and C. J. Snijders. 1995. Design criteria for the reduction of shear forces in beds and seats. *Journal of Biomechanics* 28: 225–230.

Groenesteijn, L., R. Ellegast, K. Keller, H. Berger, and P. Vink. 2010. Influences of office tasks on body dynamics using dynamic office chairs. In *Advances in occupational, social, and organizational ergonomics*, eds P. Vink and J. Kantola, (452–461). Boca Raton, FL: CRC Press.

Groenesteijn, L., P. Vink, M. de Looze, and F. Krause. 2009. Effects of differences in office chair controls, seat and backrest angle design in relation to tasks. *Applied Ergonomics* 40: 362–370.

Hamberg, H. H., A. J. van der Beek, B. M. Blatter, M. P. van der Grinten, W. van Mechelen, and P. M. Bongers. 2008. Does musculoskeletal discomfort at work predict future musculoskeletal pain? *Ergonomics* 51:634–657.

Harrison, D. D., S. O. Harrison, A. C. Croft, D. E. Harrison, and S. J. Troyanovich. 2000. Sitting biomechanics, Part II: Optimal car driver's seat and optimal driver's spinal model. *Journal of Manipulative and Physiological Therapeutics* 23 (1): 37–47.

Hazlett, R. L., and J. Benedek. 2005. Measuring the emotional reaction to passive first impression of software. In *Proceedings of the Conference Designing Pleasurable Products and Interfaces,* ed. S. Wensveen, (pp. 57–70). Technical University/Eindhoven, the Netherlands.

Hosea, T. M., S. R. Simon, J. Delatizky, M. A. Wong, and C. C. Hsieh. 1986. Myoelectric analysis of the paraspinal musculature in relation to automobile driving. *Spine* 11: 928–36.

Kroemer, K. H. E., H. B. Kroemer, and K. E. Kroemer-Elbert. 2001. *Ergonomics: How to design for ease and efficiency.* Upper Saddle River, NJ: Prentice Hall.

Kuijt-Evers, L. F. M., D. A. C. M. Commissaris, A. M. de Jong, and M. P. de Looze. 2007. Workshop: Smart design for human performance. *Proceedings of the 39th Nordic Ergonomics Society Conference,* Oct 1–3, Lysekil, Sweden. (Available in CD-Rom Format, p. 4.)

Looze, M. P. de, L. F. M. Kuijt-Evers, and J. van Dieën. 2003. Sitting comfort and discomfort and the relationships with objective measures. *Ergonomics* 46: 985–997.

Lueder, R. 2004. *Ergonomics of seated movement. A review of the scientific literature.* Encino, CA: Humanics ErgoSystems.

Mergl, C., 2006. Entwicklung eines verfahrens zur objektivierung des sitzkomforts auf auto-mobilsitzen. PhD disser., Lehrstuhl für Ergonomie, Technische Universität München.

Nathan-Roberts, D., B. Chen, G. Gscheidle, and D. Rempel. 2008. Comparisons of seated postures between office tasks. *Proceedings of HFES2008* (Human Factors and Ergonomics Society), pp. 692–696.

Nordin, M. 2004. Zusammenhang zwischen sitzen und arbeitsbedingten rückenschmerzen. In *Ergomechanics*, ed. H. J. Wilke (pp. 10–35). Aachen, Germany: Shaker Verlag.

Park, S. J., Ch. Kim, C. J. Kim, and J. W. Lee. 2000. Comfortable driving postures for Koreans. *International Journal of Industrial Ergonomics* 26 (4): 489–497.

Riseberg, J. et al. 1998. Frustrating the user on purpose. Using biosignals in a pilot study to detect the user's emotional state. *Proceedings of CHI98 v2*, New York: ACM Press, pp. 227–228.

Rosmalen, D. van, L. Groenesteijn, S. Boess, and P. Vink. 2009. Using both qualitative and quantitative types of research to design a comfortable television chair. *J. of Design Research* 8(1): 87–100.

Tierney, J. 2004. Politics on the brain? *The New York Times*, April 20.

Vercaygne-Bat, G. 2008. Specific seat requirements for market specificities. Paper presented at the IQPC Innovative Seating 2008 Congress, Frankfurt, Germany, April 20.

Vink, P. (ed.). 2005. *Comfort and design*, Boca Raton, FL: CRC Press.

Vink, P., R. Porcar-Seder, A. Page de Poso, and F. Krause. 2007. Office chairs are often not adjusted by end-users. In *Proceedings of the Human Factors and Ergonomics Society (HFES) 51st Annual Meeting*, October 1–5, Baltimore, MD. CD-ROM available.

Wilke, H. J., P. Neef, M. Caimi, T. Hoogland, L. E. Claes. 1999. New *in vivo* measurements of pressures in the intervertebral disc in daily life. *Spine* 24 (8): 755–762.

Zenk, R. 2008. Objektivierung des sitzkomforts und seine automatische anpassung, PhD disser., Technical University, München.

Zenk, R., M. M. Franz, and H. Bubb. 2011. Spine load in the context of automotive seating. Forthcoming. *Applied Ergonomics*.

5 The Ultra Comfortable Flight Experience

Overview: Previous chapters are based on facts. This chapter is fantasy inspired by the facts of the previous chapters. It consists of ideas for making the flight comfortable, which are partly realized at this point and are technically feasible. It starts with checking in at one airport and ends at the gate of your destination. The comfortable flight starts by a simple booking procedure. The booking process shows clearly what you will get because expectations do influence the comfort experience. On the departure day, the airport senses your presence and gives you specific information on how to reach the gate and tells you the waiting times on your way there. In the airplane after takeoff, you can sit in the working area where there is virtual reality (VR) or a high-quality screen available. An optimal adjustable chair, a table, good light, and temperature support you while you work. In the reading and rest area of the aircraft, the facilities are there for a relaxed flight, and in the communication area, you can phone and/or chat with passengers. There is a friendly and helpful crew, and the flight goes according to plan, and the experience is more than expected.

FIGURE 5.1   A model of the interior of a blended wing of a future airplane. Delft University of Technology.

INTRODUCTION

This is descriptive of a journey inspired by the outcome of the survey described in the previous chapters. The ideal passenger experience is the central theme of the journey. There are many future airplane concepts available. For instance, Finnair (in 2008) described the airplane of 2093—a blended wing plane. In Hamburg, the BWB AC20.30 was drafted by the University of Applied Sciences of Hamburg, and the Boeing 797 is another example of a future airplane. In Figure 5.1, the interior of a blended wing body designed by the Delft University of Technology is shown.

The focus of these future aircraft is to give an insight into the coming technological designs. The future of flight in this chapter is not on the technology of the airplane, but on the future passenger experience. Most of the technological elements needed for this comfortable experience are available, but are now only partly implemented. In the description of the flight experience, the flight starts at home and ends at the arrival of the airport.

THE FLIGHT EXPERIENCE

AT HOME

The ideal comfort flight starts at home or at work. After login on the Web site of your favourite airline, you only have to add the destination and date. It is not necessary to add other information. However, if you have special requests, it is possible to easily add your specific preferences. The airline has assured you that the system is very safe and stable and always works correctly. Your identity is registered by putting your fingers on a device or you just have to turn your Webcam on and, by face recognition, your personal data pop up immediately. If you have special requests, such as another departure airport than one you usually use, it is possible to easily add your specific preferences. Several alternatives are shown with the comfort ratings of other clients. Also, two or three positive and negative comments are shown. When you affirm the booking, you see the interior of the plane on your screen. This is not the drawn seat map with seat numbers, but a three dimensional visualization. You see the interior from the position where you enter the plane to give you the right expectations as you walk around in the virtual plane. You select a seat on the seat map, and you see the

FIGURE 5.2   One of the views you see on the Internet while booking your seat, which influences your expectations for the flight.

interior from the position of your selected seat (Figure 5.2). This is important, as expectations influence the comfort feeling during the flight. If you want, you also can look down, enter your body height and see an estimation of what your space is around your legs.

On the seat map, you will see different areas in the plane (Figure 5.3):

- A working area, where it is quiet with enough space for a laptop (or other device) and power supply for the laptop in existing planes. In new airplanes you can use just your memory stick or other memory devices in the supplied laptops. Perhaps you work with cloud computing where your personal data are available via a kind of Internet system. This system is also available in the airplane. It is also possible to use systems like a cell phone (which is only able to make text messages or Internet connections to reduce disturbance of your neighbours) or a personal digital assistant (PDA). A memory device is also enough, as the large screen in your seat shows the information and a keyboard is available on your table. It is also possible to wear a virtual reality device by which privacy is ensured (Figure 5.3). A light for your book and an extra table for documents are arranged as well. The seat, keyboard, and screen can be adjusted to be able to work in a comfortable position, similar to your on-the-ground working environment. The food is positioned and brought in such a way that you can continue to work.
- A communication area where you can use your cell phone. Seats can be turned within the pitch to enable you to have face-to-face meetings with your friends. Also, you have access to your personal data on the Internet or that you can access via your cloud computing system.

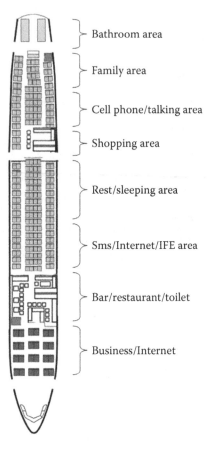

FIGURE 5.3 A seat map with different areas allocated to support the activities the passenger wants to do during the flight.

- A family area where children can play IFE games and parents can take care of the youngest. The children can put their picture in the system and can play against each other (or other passengers) with the virtual reality system (see Figure 5.4) and see avatars of each other in the games.
- A rest area where you can read or sleep with the legs off the ground and your head and neck supported. It is also possible to sit sleeping sideward. You are only disturbed for food if you prefer.
- An environment experience area where you can look through the window and outside via cameras and gather extra information on where you fly (Figure 5.5).
- A low-cost area for the cheapest seats with no extra amenities.

You can book an eight-hour flight and book the work area for the first two hours, the sleeping area for the next five, and the communication area for the last hour (Figure 5.5).

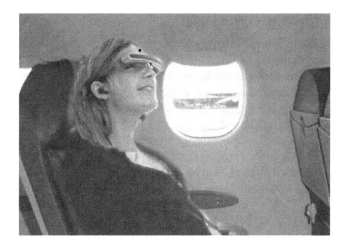

FIGURE 5.4 A virtual reality device, which affords more privacy while working with your documents or making it possible to play three-dimensional games and movies. It also saves weight and space.

FIGURE 5.5 Information on the airplane ceiling on the location of the plane while in the air (Haperen et al. 2005).

FORTY-EIGHT HOURS BEFORE THE FLIGHT

You will receive an e-mail 48 hours before the flight directing you to print your boarding card with merely two clicks of the mouse. There is no need to bother about your seat assignment as it was already selected while booking. There is an option to get advice on a route to the airport and a map of the airport with the way to the gate. For different obstacles (baggage drop off, check-in, customs, and gate), the needed documents and activities are added to the map, with pictures. This also is added for the return flight because, on a holiday or work destination, printing facilities often are not available. Of course, in the future, all airports will have facilities to read your fingerprints, and boarding cards will not be needed, but that will take a few years to come to fruition in all areas in the world.

TO THE AIRPORT

Because you have flown before and your preferences are in the system, you will have advice on the best way to get to the airport (train, taxi, car, bus, or shuttle). You follow the advice and arrive at the airport. To navigate to the airplane, the advice is printed or integrated into your mobile phone or blackberry.

AT THE AIRPORT

The environment at the airport is well designed and makes you feel comfortable. Plants, wood, and nice views give you the feeling of being welcome. It is not like in Figure 5.6. The sensors at the airport record your position and your personal data and convert this into useful information for you. Arrows appear on the ceiling and

FIGURE 5.6   You do not feel welcome in a cold atmosphere.

wall, or the mobile phone talks you through the first obstacle: the luggage drop off. There is no line for the luggage drop off where the signs have directed you. You can put the suitcase on the conveyer belt without lifting the suitcase. It recognizes you and asks you whether the destination is according to your ticket. This gives you the feeling that the luggage will arrive safely at the correct location. You push the button "yes" and the system attaches this destination automatically to your suitcase. At the airport, the signs now show you clearly where you have to go. Then you arrive at the next obstacle: the security check. This time the security agents are trained to not feel important or to be rude (sometimes it looks like they all have to be arrogant, uninterested in their job, and unpleasant). No, this time they are correct, precise, and friendly and give you a safe feeling. They are trained to be flexible and to make sure that there are no passenger lines. You do not have to take your laptop and fluids out; you do not have to lift up your hand luggage. You place it into a system that picks it gently up and down again. After checking, it is placed in a position where you can pick it up easily. The systems are able to do the body scan while you put your hand luggage in the system. So, there is no time loss here. At customs, you know which documents you will have to show because it is shown on the wall. As a frequent flyer, you can be guided by the signs on the wall and ceiling and by your mobile phone, and if you want to go the restroom, you just say that to your cell phone and guides you there.

THE LOUNGE

The business lounge is an oasis of rest. You are assisted again in a friendly way and your favourite drinks and snacks are there as well. The lounge chairs are very comfortable. Studies described in the previous chapters have shown (see Figure 5.4) that variation of posture is observed frequently in the ideal comfort situation at home. Therefore, this chair accommodates a variety of postures (Figure 5.7): for watching a television screen, reading, listening to music, or working. The colours, air quality, and lighting are providing for wonderful relaxation. Your flight and gate number have already been marked on the screens and the walking time to the gate is shown as well. The boarding time is shown and the nice voice of the ground staff is telling you to go to the gate because the boarding time is almost there. The signs are again very clear on the wall and ceiling and you know where to go and have the feeling everything is okay. In case this system does not work, you also receive a map with your personal route. It shows the route along your favourite shops as well as a direct one.

AIRPORT PLUS

In some airports, you are treated like royalty. Your mobile phone or personal tag (which could be on your frequent flyer card or it was sent by mail some days before the flight) is connected to the environment. On the first message board, you are personally welcomed by a message. Hereafter, projected signs on the floor guide you through the building to the various locations where you have or want to be. They also show pictures of people you may know. If you click these, they tell you where the persons

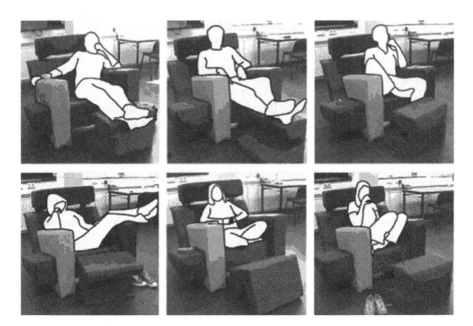

FIGURE 5.7   Some postures observed in a comfortable seat, which is made to support various positions in a comfortable way. (Design from D. van Rosmalen.)

are. However, these persons must first give permission. You also are asked by your mobile phone if you give permission for the system to show your position to others.

Some people might not like the new electronic systems, the signs on the wall, or on the mobile phone. In these cases, a friendly person comes to you when you enter the airport building and shows you on a map where you should go and which documents are needed where. You can ask for assistance anytime by calling the special number on your mobile phone or placing your hand on the wall (the gestural recognition system senses this).

In case you have more time to spend at the airport due to a delay or other reasons, there are not only shops, but other entertainment as well in special dedicated areas. In these areas, a museum with special art or culture of the country could be included or it could be an area with a virtual environment where you name the places where you have been during the trip and movies will be shown of those places. It could also be a miniature city that offers you the highlights of a country (Figure 5.8). You can see parts of the country in miniature that you have seen in reality. These areas make you experience the environment again in another way. There could also be an area where a live band or video is playing the local music and local smells are distributed as well.

AT THE GATE

At the gate, you are again assisted in a very friendly way. You can sit in seats that give you a good view of the gate and the plane you are about to board. This plane looks spectacular and attractive (see Figure 5.9). It is not a dark or blind wall, but there is a

FIGURE 5.8   A miniature of the city that you just visited could make you reexperience the environment.

good view. All cars and trucks at the airport are electrically driven, which is good for the environment. There is no bad smell. The airport is the ideal place for electric cars and trucks, as the reach distances of electric cars are restricted and at night there is time for battery loading. Some airports are already experimenting with these systems. The plane is boarded by two entries: in the rear for the back seats and in the front for the front seats to reduce waiting times during boarding. For wide bodies even more entries are used.

You are able to experience the interior of the airplanes that used to fly on this route and the current seat. The old and new seats are positioned at the gate and you can try these. This way you become aware of the improvements. You can compare the foam of the seat and experience the improved leg space and the roominess.

THE PLANE ENTRANCE

The entrance of the plane is a nice experience as well. You do not see the old aluminium industrial look (Figure 5.10), but a nicely illuminated cabin interior having the home interior atmosphere. The jet bridge has a good temperature and a good view. You do not have to wait in a hot (or freezing) jet bridge because there are two boarding entrances to the plane and the front half of the plane boards first (the other half is boarded by the other jet bridge). You also can walk directly to your seat without waiting for the other passengers because the boarding process is well organized. The crew welcomes you on board with a friendly welcome, you can pick up a newspaper on your way to your seat, and the plane picks up your mobile phone or personal tag and guides you (with the aid of lights) to your seat. All passengers are well instructed to limit the size of the hand luggage in the plane. This way you can stow your hand

FIGURE 5.9 An airplane draped in the colours of the state of Maryland (U.S.) arriving at your gate, and an airplane showing a special event (Asian Games in Doha, Qatar), both of which are nice to see when they arrive at your gate.

FIGURE 5.10 The often seen entrance of an airplane, which has many possibilities for improvement. It is like a welcoming through the garage.

luggage with ease. In your seat area, you have the possibility to store the things you need during the flight. As was described above, each area has its own atmosphere. The work area has more blue light and you can arrange your preferred light, such as the special white light, which stimulates your alertness. This lighting system is now used after lunch in some offices to reduce the postlunch dip (Vink et al., 2009). If you want to feel more relaxed you can turn on the warm light. In the sleeping area, warm light is used, while in the communication and family area, warm red light is available. The colours of the areas also make you recognize your seat. Your seat looks comfortable, and the area where you will sit is illuminated giving it a bright and roominess look and giving you the feeling of being very welcome.

The temperature of the plane is 18°C (64°F) and you have the possibility to adjust the temperature of your feet, back, and neck locally and by ventilation. This is not an extraordinary feature. Many studies on temperature show that having control of your own temperature increases the comfort experience (e.g., Bordass and Leamon,

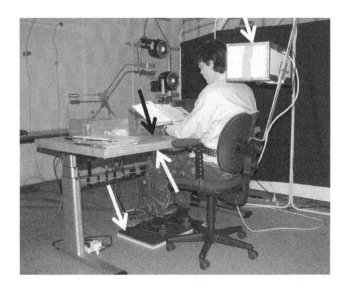

FIGURE 5.11   Studying the effect of local heating systems on comfort (Oeffelen et al., 2009). The arrows are local heating sources.

1997). In offices, it saves energy to have the air temperature 18°C and where it is possible to add some local personal heating.

A study of Oeffelen, Zundert, and Jacobs (2009) showed that temperature regulated locally increases the comfort and reduces energy consumption (Figure 5.11). It is worthwhile studying if lightweight local heating systems now under development for electrical cars (Canton, 2010) are usable in aircrafts. The screen shows the real feel temperature of 18°C on the three areas on a human body, and, per area, an arrow upwards; if it is touched, a higher temperature is indicated and realized. The crew is advised by the system that there are new passengers who have not flown this type of plane. They aid the passengers in using the system to adapt the temperature to the passengers' preference.

The Long Haul Flight

The seat depth and armrest are long enough to support you with comfort, but short enough to make easy in- and egress possible and give space for leg movement while seated. The flight attendant is attentive for your need to rest, to work, use IFE, or comfort the children. Additionally, you can choose two pillows (Figure 5.12) from an assortment to vary your posture for a sleeping or resting position. You are offered a 10-minute foot massage if you want. The crew has a product that can be placed over your socks, and it provides a relaxing movement around the feet and lower legs. It gives you a superb comfortable feeling. Roominess of the interior and a good view outside the plane are taken care of as well. You can see at your screen or on the ceiling (see Figure 5.5) the environment and zoom in on areas that have your interest. You can see the stadium in the city you pass and watch the match in that stadium or you can see some typical fishing boats of the area and fish in the

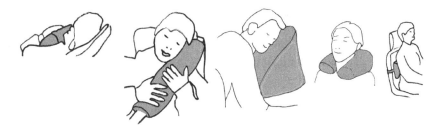

FIGURE 5.12 Several forms of pillows that can be given to passengers taking into account individual wishes for the passenger's comfort and facilitating the change in postures.

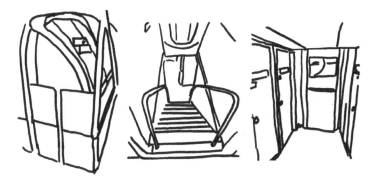

FIGURE 5.13 A toilet downstairs (left: the entry, middle: stairs, right: the toilets) reduces odours and stimulates extra movement.

sea below you. As described before, you can adapt the temperature in the floor and around your head and have access to some of the hand luggage while seated. The hand luggage is stowed above your head to give more movement space around your legs. By a push on the button, the hand luggage above your head comes within reach without disturbing other passengers. The toilets are not close to you, so you do not have the bad odor. Most toilets are downstairs (Figure 5.13), which is good to help stimulate some physical exercise. There is one toilet available for persons who are physically less able to climb the stairs (there are also special seats for disabled persons and children).

Taxiing to the runway is done by an electric vehicle and starts soon after embarking. Before reaching the runway, the engines are started and at the runway there is no queue. The planning is made in such a way that the plane is immediately cleared for takeoff. (There are times now that you feel that in the traffic tower everyone is surprised and thinking: "Never thought that so many planes would be departing and now another a plane is coming, oh, man." Or they think: "Hey guys, let's make a nice queue of 10 planes.")

During the flight, the messages of the crew are clearly given in your native language. This seems a normal situation, but many passengers complain that the messages are difficult to hear because of bad English or because the system is not used properly. (Sometimes it sounds as though some pilots put microphones deep in their throat while talking.)

In the sleeping and working areas, your food is served when you request it. The food is not the normally strange aircraft food. Nowadays, aircraft food is usually just bearable to eat. No, this time the airline has studied which food is most appreciated by passengers. There is a bottle of water already in your seat and the cold drinks can be taken at the bar.

BUSINESS CLASS

In business class, you are welcomed in a unique way. High-quality lounge music, a superb design, and special luxury lighting is attracting you. You are personally welcomed by the crew and guided to your seat. It is easy to stow your hand luggage. The luggage bins come down at your elbow height, and you can place hand luggage in the bins easily. The cushion and blanket are not on your seat. Nowadays, often you have to put away your own stuff and the stuff that is already there in your seat: the cushion and the blanket (Figure 5.14). Putting away the cushion is difficult as you do not want to put it on the ground. Later you will be sleeping on this cushion. Now, when you arrive at your seat, the seat is free. Also, in the upright takeoff position, the seat is comfortable, which is now often not the case in business class. Adjusting the seat is easy. It has a button ("ideal seat"), such as in the experimental BMW 7-series (Zenk, 2008). The seat records the pressure between the seat and your body and calculates the ideal pressure distribution and adapts itself. In the sleeping position, the seat is a flat bed. This is definitely not the now often-seen flat bed with many hard and soft parts, and not the forward rotated flat bed having the effect that you glide off the seat. No, the seat has a form that follows your body contour. The intelligent seat again adapts itself to the ideal rest position by using air cushions preventing you

FIGURE 5.14   Arriving at your seat confronts you with the first problem: Where to stow the cushion?

from feeling the mechanical structures. Your head and neck rest also can be adapted by simply bowing parts of the seat. The neck rest has soft foam and the headrest somewhat harder, as studied by Franz (2010). A blanket can be put over the headrest. The form makes you feel comfortable, and there is no need for an extra cushion. However, there is an extra cushion for pure comfort. The seat also records the time you are in one static position. If it is too long, the seat adapts itself or asks you if it should change its position to make you feel more comfortable. There is a tested massage system, such as the lightweight massage system, in Franz's study. The crew asks you about your preference regarding food, drinks, sleeping time, IFE use, VR use, reading, and working time, and adapts their schema to your wish. There is no neighbour, and you are free to go to the bar and buffet whenever you wish without disturbing anyone. The seat has an exercise system in it to make you feel more fit at the end of the journey. Straps can be placed around ankles and wrists. You can push your shoulder in the backrests and sensors record your force. You can follow an interesting instruction program especially designed to feel fit after a flight or you can play games with the VR glasses mounted on your head. It is made in such a way that it does not disturb your neighbour.

Inexpensive Flight

For the inexpensive flight, ultrathin, lightweight seats are used and even lower-priced halfstanding seats can be booked. Halfstanding is a position you sometimes take in a bar using a stool. Twenty minutes after takeoff drinks and food can be bought in the shop in front of the plane and tax-free gifts in the back of the plane. This movement makes you feel more fit after a plane trip. You can see the gifts and, after tax-free ordering, they will be sent to your home or holiday address. The flight attendants will not block the path with their trolley as the food and products are in the shops. The seats are not adjustable making it more lightweight and preventing discomfort for the passengers behind the one adjusting the seat. Halfway through the flight, the half-standing and the sitting passengers change places. Studies of Vink et al. (2008; 2009) show that alternating between sitting and standing is more comfortable. Research is needed, however, regarding the optimal halfstanding position in the airplane.

In-Flight Entertainment (IFE)

As was mentioned above, your virtual reality system or high-quality screen and high-quality sound is working perfectly. The airlines check these systems regularly. You can watch your movie or live television programme at a time of your choice and in your language. It is easy to control the system. For every handling, you get immediate feedback, which gives you the feeling that you are in control of this system.

Cleanliness

The airplane has a bright and new looking interior, and it is easy to clean. Broken elements can be replaced easily in between flights by a click system and special cheap systems are developed to clean the plane often.

CREW

The crew is selected and trained to comfort passengers. That means that the crew is incredibly interested in people and has a caring attitude. It sounds like nothing special, but, if we read the trip reports, many passengers are disturbed by the attitude of the crew. It is better if a pilot says nothing than hearing the voice of someone that is not interested in whether or not his message is received by passengers because of a bad tuned system or a bad language pronunciation. A pilot should be concerned that his or her passengers have a comfortable flight and should feel safe and well informed about unusual feelings. Pilots are trained to know what passengers experience as unusual. A flight attendant asking "tjik or chee" of passengers, meaning "would you like chicken or cheese on the sandwich," is missing one of the few opportunities to have contact with the passengers. The pilots and flight attendants inform you in the beginning of the flight on the safety by the intercom. All other information will be given by text on the IFE, giving the possibility for passengers to choose whether they want information on the location in the air and if something special happens in the plane, or whether they want rest and no disturbances. If the crew is asked to adjust something like temperature or food, the crew gives clear and friendly information, whether the adjustment will be done or not. It is better to clearly explain why some things are impossible than to neglect to point this out to a passenger.

In case a passenger has some fear for flying, the crew is attentive and special information will be given. The reason for the different sounds are shown on the screen. For instance, in the middle part of an Airbus A320 series, noise can be heard by the passengers, which is from the engines in the fuselage that are connected to parts of the wing like the flaps. In those seats, an explanation will be given on the screen. Also, other noises are explained. For instance, after climbing, the engines will reduce their noise, which will be explained.

ARRIVAL

The landing is announced and specific information for your connecting flights, gates, and route description are given on the screen for every passenger. If the passenger has changed his seat, he can adjust the seat to his name and will get the proper information. The route description can lead to the rental car companies, railway station, or area where friends can pick you up. Also, information is given regarding the estimated passport control time and luggage waiting time. The airport expects the airplane. This also may sound silly, but many trip reports mention that planes are instructed to hold in the air or on the ground. On the ground, the jet ways are sometimes not available or the staff controlling the jet ways is missing. Sometimes push-back cars are missing or gates are occupied. This is strange because the airplane screen often shows exactly the number of minutes before landing, but on the ground everyone seems surprised that a plane is coming in. Is it possible that no one at the airport has access to the number of minutes before the plane arrives? In the trip reports, most passengers prefer a jet way before a bus. The bus is often to hot or too drafty or waiting for the other passengers, which is seen as very uncomfortable. An airport seems a strange business to passengers as sometimes the airplane is parked in front of the terminal and yet a bus still has to be used to bring the passengers to the terminal. However, in this comfortable flight, the jet ways are used and the plane is expected.

At least two doors will be opened—one in the back and one in the front of the plane—to increase debarking time. The hand luggage is easy to reach. Armrests of chairs are short or go up making it easy to leave the plane with bags and friendly flight attendants guide you to your next part of the journey. From the staff, you hear why a possible discomfort was a coincidence in this flight, and they reiterate the normal situations for this aircraft. Again, you are picked up by the sensors of the airport, and the mobile phone or signs at the airport lead you clearly to your destination. You have had a very comfortable flight, and you are ready for your next activity. You were not aware of the smart lighting system, but this light in combination with the exercises in the seat make you feel extra fit and ready for your new activity, and you have only good memories of this trip.

REFERENCES

Bordass, W., and A. Leaman. 1997. Strategic issue in briefing, design, and operation. Future buildings and their services. Strategic considerations for designers and clients. *Building Research and Information* 25 (4): 190–195.

Canton, J. 2010. Comfort by lightweight temperature systems in electric cars. MSc thesis, Technical University, Delft, the Netherlands.

Franz, M. 2010. Comfort, experience, physiology, and car seat innovation. PhD thesis, Technical University, Delft, the Netherlands.

Haperen, van , B. J. A., P. Vink, C. J., Overbeeke, J. P., Djajadiningrat, and S. H. Lee. 2005. Concept of a future Nissan car interior. In: P. Vink. *Comfort and Design: Principles and Good Practices.* Boca Raton, FL: CRC Press, 263–280.

Oeffelen, E. C. M. van, K. Zundert, and P. Jacobs. 2009. Local climate influence. A lab study with subjects to the possibility of local body temperature regulation (in Dutch). Delft, the Netherlands: TNO Bouw en Ondergrond.

Vink, P., L. Groenesteijn, M. M. Blok, and M. den Hengst. 2008. Effects of a meeting table and chairs making half standing possible. In *Conference Proceedings, 2nd International Conference on Applied Human Factors and Ergonomics*, eds. W. Karwowoski and G. Salvendy. July 14–17, Las Vegas, NV. Louisville, KY: AHFE International, CD Rom.

Vink, P., I. Konijn, B. Jongejan, and M. Berger. 2009. Varying the office work posture between standing, half-standing, and sitting results in less discomfort. In *Ergonomics and health aspects of working with computers*, ed. B. Karsh (pp. 115–120). Proceedings of the HCII2009 Congress, San Diego, CA, July 19–24. Berlin/Heidelberg: Springer Verlag.

Zenk, R. 2008. Objektivierung des sitzkomforts und seine automatische anpassung. PhD thesis, Technical University, München.

6 Illustrations and Comments on Aircraft Interior Comfort and Design

Overview: In Chapter 3, 10,032 trip reports of passengers, who flew in 2009, were analysed. The factors that have the most influence on comfort are leg room, hygiene, and crew, but the influence of the preflight experience, seat, and luggage is substantial as well. In this chapter, 59 photos taken by passengers illustrate somewhat the analysis in Chapter 3 as well as some new points shown below.

INTRODUCTION

In this chapter, 50 photographs taken by the passengers illustrate the points of the analysis in Chapter 3. The issue of leg room is well illustrated, but some points are somewhat more difficult to show in a photo, such as service and crew attention, and, therefore, are undervalued in this chapter. Some photos concern issues not mentioned in the studies, but are shown as a source for further consideration. The original text of the passengers is added to the photos. Some photos taken by the passengers are already used as illustrations in Chapters 1 through 5 and will not be repeated here.

LEG ROOM

FIGURE 6.1   "This is what I call leg room. It's more than I ever had in the economy class. I tried to reach the seat in front of me, but didn't succeed."

FIGURE 6.2   "It does not make sense to position the seat pocket where I need [the] most leg room."

FIGURE 6.3   "I need much time to recover from this position. In some planes, it is even worse as window-side rows have entertainment boxes underneath seats that adversely affect legroom."

FIGURE 6.4   "Do you see the third row? That's where I want to sit for the leg room."

FIGURE 6.5   "This is what I call leg space in an airplane."

FIGURE 6.6   "A warm, smiling flight attendant makes you feel welcome."

SERVICE

FIGURE 6.7   "Newspapers in the economy class is nice service."

HYGIENE

FIGURE 6.8   "A spotless clean, nice smelling toilet makes your trip pleasurable."

FIGURE 6.9   "If the stickers are like this (loose), what would the engines and cockpit look like?"

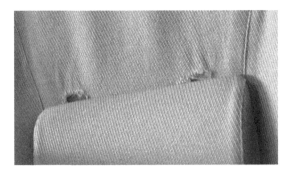

FIGURE 6.10   "This worn-out business class seat does not make a luxury impression."

FIGURE 6.11   "A broken tray table? Or is this the normal situation in the business class? It is very unstable to do your work."

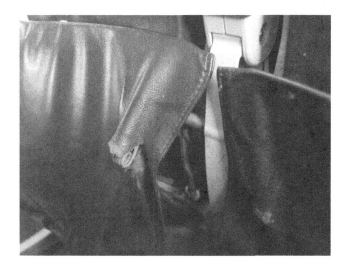

FIGURE 6.12   "This seat part looks likes it needs some maintenance."

LUGGAGE ROOM

FIGURE 6.13   "Where do I put my carry on luggage? Under the seat will affect my legroom."

FIGURE 6.14   "Hard to reach the luggage bins."

NEIGHBOUR

FIGURE 6.15 "Three seats left and three seats right. Other airlines have a 3-2 seat configuration in this Bae146, which much more comfortable."

FIGURE 6.16 "This Boeing 777 has a 3-4-3 seats configuration, while other 777s have a 3-3-3 in the same airplane, which is more comfortable, of course."

FIGURE 6.17 "The single seat in this E145 is not so bad. You have no neighbours."

FIGURE 6.18 "The passenger close to this lady had the ventilation on in the wrong direction."

SEAT

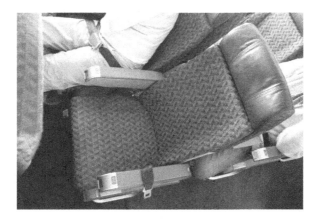

FIGURE 6.19   "The seat pan is too short for a two-hour flight."

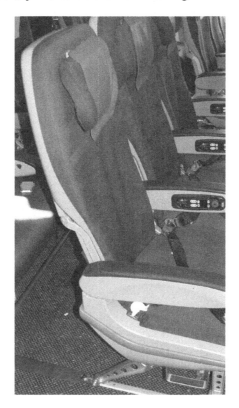

FIGURE 6.20   "This seat has a nice thin backrest at knee level, but the seat pan is somewhat short."

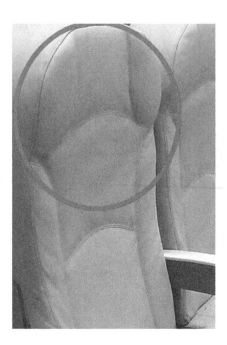

FIGURE 6.21   "This seat has no free shoulder space (I am a biomechanical specialist), which is important also for comfort."

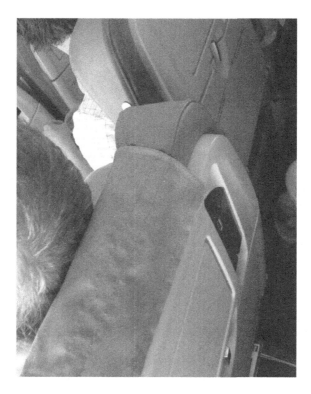

FIGURE 6.22   "It's a disaster if the person in the seat in front of me is pushing the button in the armrest. My knees damage, I can't eat anymore and I have no living space. Kill the inventor of the button."

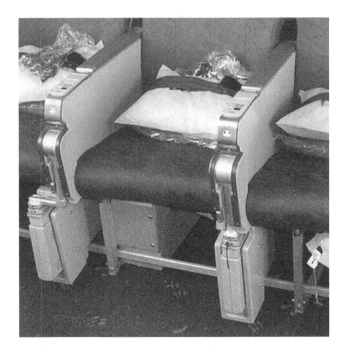

FIGURE 6.23 "The seat width is drastically reduced in the front row, because someone had the idea that a tray table in the armrest is possible. Is there no better solution?"

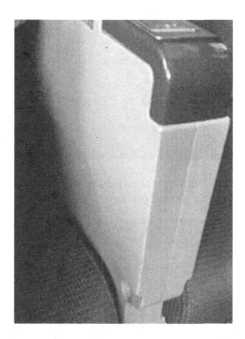

FIGURE 6.24   "The seat width is reduced by the arm rest. I have had flights where the flight attendant takes a tray table out of a luggage bin, which is much better."

FIGURE 6.25   "Thick cushions look more comfortable."

FIGURE 6.26 Compared with figure 6.27, a passenger mentions that this position is wrong as the buttons on the side make you step out of your seat to see and reach the buttons.

FIGURE 6.27 A passenger mentions that buttons on the top of the armrest are better, visible, and easy to touch compared with buttons on the inner side.

FIGURE 6.28   "I would like to sleep in this position, but the seat is not really supporting this."

FIGURE 6.29   "The armrest is too low."

FIGURE 6.30 "This is why I prefer a window seat."

IFE

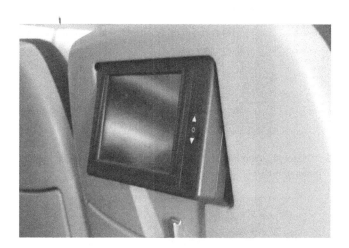

FIGURE 6.31   "It's important to have a screen that can rotate like this one. I have had screens fixed in the seat, which makes the viewing angle terrible."

FIGURE 6.32   "This is a nice large screen."

FIGURE 6.33   "Safety instructions by the screen are much better. As it is often better to see, to hear, and to understand."

FIGURE 6.34   "In the middle, there are four buttons (circle) that do not make any sense. It's nice that you can have your arm on the arm rest without touching the control."

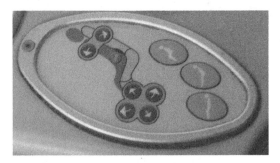

FIGURE 6.35   "This interface is not that bad."

FIGURE 6.36   "This control cannot be reached sitting in the chair."

DELAY/WAITING

FIGURE 6.37   "I hate long waiting after landing. Probably everyone at the airport was surprised that again a plane landed."

FIGURE 6.38   "Always long walking distances at airports."

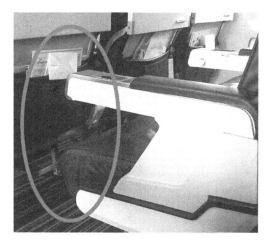

FIGURE 6.39   "Long arm rests block in- and egress and give nothing extra."

OTHER ILLUSTRATIONS

Apart from the issues mentioned in Chapter 3, additional photos were submitted on the design, safety, and on business class comfort. These do not have the highest priority in reference to the outcomes of the study, but might be interesting for readers and, thus, are included below.

DESIGN

FIGURE 6.40   "This is not a warm entry."

FIGURE 6.41   "This is what I see from my seat. Is a designer involved or did no one think about what passengers do see from their viewpoint?

FIGURE 6.42 "I booked a business class seat and this is the stunning view: white panels."

FIGURE 6.43 "What a mess in the business class."

FIGURE 6.44 "Seats with different colours and texts, the interior did not really make a calm impression on me."

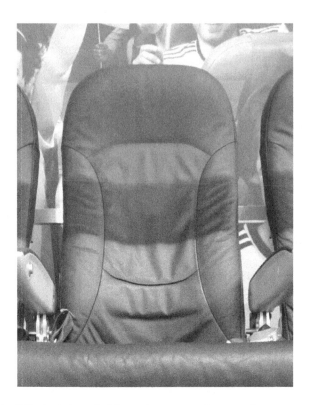

FIGURE 6.45 "The background of the seat was a surprise to me."

FIGURE 6.46 "Who designs the position of the windows?"

SAFETY

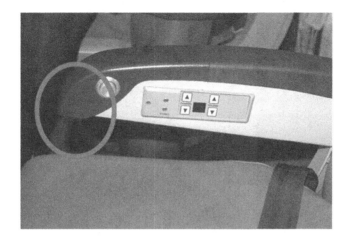

FIGURE 6.47   "Sharp edges."

FIGURE 6.48   "I scratched my shin and the pantyhose was damaged."

FIGURE 6.49 "The sharp edge is pointing in my thigh."

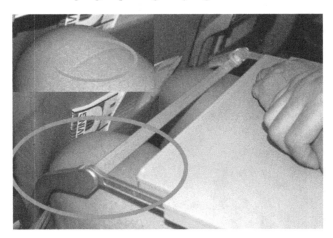

FIGURE 6.50 "My leg hurts."

BUSINESS CLASS

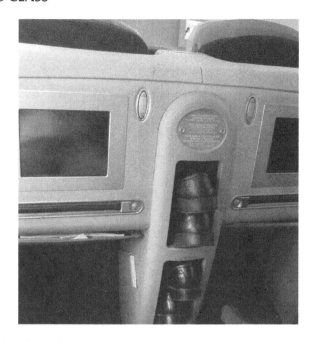

FIGURE 6.51   "Good idea to stow you shoes."

FIGURE 6.52   "Warm-coloured business class."

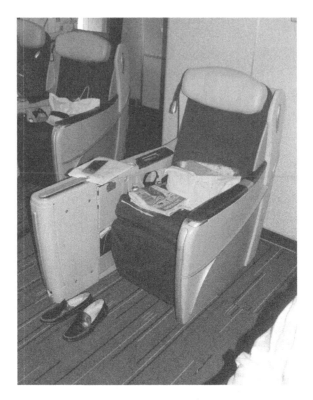

FIGURE 6.53 "This seat is without neighbours. You can leave the seat whenever you want, which is not so bad."

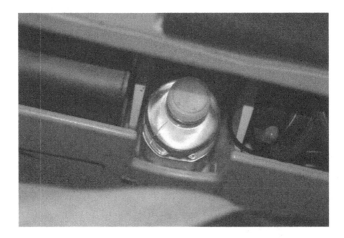

FIGURE 6.54 "Nice to have a good location for your water."

FIGURE 6.55 "You cannot sleep or relax in this business class."

FIGURE 6.56 "Some airlines call this a business class."

FIGURE 6.57 "Some airlines call this a flat bed, but it has many hard elements and often flat beds are not horizontal, which is better than gliding out of the flat bed."

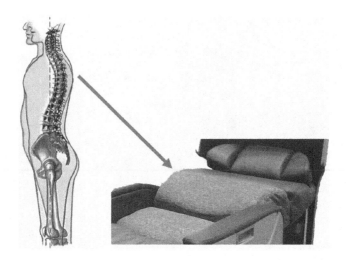

FIGURE 6.58 "First study the human anatomy and then design the seat," comments a physiotherapist.

FIGURE 6.59 "Nice first class design, with warm colours."

Book Summary

This book is important for those working in the aircraft interior industry because it contains important information about theories on comfort gathered directly from the voice of the passenger. The research is based on reactions from 10,032 passengers in 2009, which includes specific tips and photographs on passenger likes and dislikes as well as an overview on the latest scientific demands for passenger seats.

Index

.

Printed in the United States
by Baker & Taylor Publisher Services